Ornamental Fish Culture
Aquarium Manageme

Ornamental Fish Culture and Aquarium Management

Dr. A.D. DHOLAKIA

M.Sc., LL.B. (Sp), Ph. D.
Research Officer and Head
Fisheries Research Station,
Junagadh Agricultural University, Sikka

2016

Daya Publishing House®

A Division of

Astral International Pvt. Ltd.

New Delhi - 110 002

© 2009 A.D. DHOLAKIA (B. 1947–   )
First Published, 2009
Reprinted, 2016

ISBN: 9789351241324 (International Edition)

Published by   : **Daya Publishing House®**
                         A Division of
                         Astral International Pvt. Ltd.
                         – ISO 9001:2008 Certified Company –
                         4760-61/23, Ansari Road, Darya Ganj
                         New Delhi-110 002
                         Ph. 011-43549197, 23278134
                         E-mail: info@astralint.com
                         Website: www.astralint.com

Laser Typesetting   : **Classic Computer Services**
                         Delhi - 110 035

Printed at   : **Replika Press Pvt. Ltd.**

Acknowledgement

My first inspiration to write this book was my daughter Ushma and son Apar. When they were kids, they had good attraction to see aquarium fishes. Now my daughter-in-law Kshiti also has a fond of keeping aquarium in house. My grand son Kathan is always busy in observing each movement of fish in my aquarium. I can not forget the inspiration given by my B.F.Sc. students saying that my teaching should be converted in to book. Loving affection blended with emotions and charm has constantly encouraged me to write this book.

I must thank Dr. L. L. Sharma, Director, Winter School, ICAR, College of Fisheries, Udaipur, Rajasthan for guidance and appreciating my effort saying that my knowledge should be given to employment creating entrepreneurs, culturists and to students.

I am highly thankful to my Vice Chancellor, Junagadh Agricultural University Dr. B. K. Kikani for writing "Foreword" for this book. I am also thankful to Shri K. P. Joshipura, Registrar of our Junagadh Agricultural University for writing "Recommendation" for this book.

I dedicate this book to my wife Smt Sangita to whom I received utmost encouragement and appreciation. But for the help received from her to keep myself free from any domestic chore whatsoever, the writing of the book would not have been impossible. I also express my gratitude to my late mother Smt Jayshriben and father Shri Dwijendraray.

Dr. A.D. DHOLAKIA

Foreword

One of our scientists Dr. A.D. Dholakia has written one book of *"Ornamental Fish Culture and Aquarium Management"*. At present he is working as Research Officer and Head of Fisheries Research Station of our University at Sikka.

Wind and water are two powerful forces of nature. Fish is generally regarded as one of the symbol of wealth and good fortune. If you are arranging your aquarium in your house, office or factory in right direction, it brings stream of money to you.

Considering above, Dr. Dholakia in this book has explained how to fabricate aquarium, how to arrange, which are, different types of plants can be useful in aquarium, details of more than 140 freshwater and more than 25 marine water fish with classification, identification, water quality parameters, for each of above fishes, their breeding methods, their diseases and curing methods and many more.

Being a tropical country ornamental fish culture is economically viable. As maintaining an ornamental fish farm (both hatchery and grow out) is easy and breeding can be done throughout the year, it provides good opportunity for self-employment and commercial business. It requires less initial investment compared to prawn farming.

I feel that this book will be a boon to University students, researcher, aquarium fish hobbyists and to the trade who want to earn money with less investment.

I wish that from the information of this book more people will start their own business. I congratulate Dr. Dholakia for this special type employment oriented book.

B.K. Kikani

Vice-Chancellor
Junagadh Agricultural University
Junagadh – 362 001, Gujarat

Preface

India's share in this global aquarium fish trade is only 0.007 per cent, in term of money, it comes to Rs. 23 million or 2 crores and thirty lakhs. However, India has a domestic market of Rs. 10 crores in Ornamental fish trade and this is fast growing. MPEDA (Marine Products Export Development Authority) has estimated that India's potential for trade in aquarium fish is 5 billion US dollars which we can earn by the export of ornamental fishes. "Nabard" has visualized in 2004 the rise in our trade up to 1 per cent global trade in five years.

About 85 per cent of ornamental fish species sold in the biggest ornamental fish market of India at Kolkata–locally called "Hatibagan Haat" are from wild collections in the rivers, springs nallas, ponds direct waters in the state of West Bengal, Assam, Meghalaya, Arunachal and Manipur. Only 15 per cent are the exotic fishes breed in captivity. In the eastern part of Rajasthan especially around Bharatpur emphasis is on wild fishes. The nearest markets for Rajasthan are Agra, Delhi and Jaipur.

The other two centers *viz.* Mumbai and Chennai lay emphasis on breeding of exotic varieties of ornamental fishes. This is because of the ideal climatic conditions.

Ornamental fish trade is smaller proportion also exist in Himachal Pradesh that has larger water resources in the form of rapids, ponds and tanks in the hilly region. From here wild fish varieties are collected and sold in Shimla, Chandigarh, Delhi etc.

Looking to this market availability, one can start his own culture system. By proper management one can earn good amount. In this book identification, classification of some of the important more than 140 freshwater ornamental fish and 25 marine water ornamental fish with their breeding method, culture method, aquarium behaviour and management is given. Culture method of food to be given at different stages of life is also given.

I hope, my effort will help aquaculturists, hobbyists and who are interested in aquarium fish trade, this will also be useful guidance for students.

Dr. A.D. DHOLAKIA

.

Recommendation

I am happy to note that Dr. A.D. Dholakia, Research Officer and Head, Fisheries Research Station, Junagadh Agricultural University, Sikka has written a book on *"Ornamental Fish Culture and Aquarium Management"*.

I have seen details of different aspects covered in this book. He has covered fabrication and arrangement of aquarium, its accessories, aquarium plants and its culture, methods, suitable feed both artificial and live and its culture methods.

Important thing he has explained is breeding methods in details for each fish species for more than 130 fresh water ornamental fishes and more than 24 marine water fishes their diseases and its treatment. He has also given the list of available ornamental fishes, which are available from wild in different states of India. This will be helpful to collect brood stock from natural resources.

The ornamental fish keeping, which started as a hobby now supports a commercial aquarium fish and related accessory supply business and has taken the shape of an industry.

In this respect, I feel that the work done by Dr. Dholakia is very good to University students, research workers and especially to new entrepreneurs. I, therefore, recommend that who wanted to start aquarium fish culture as employment created business should study this book and take maximum advantages from the information given.

K.P. Joshipura

Registrar
Junagadh Agricultural University
Junagadh – 362 001, Gujarat

Contents-cum-Detailed Index

Chapter 1
Introduction

Aquarium keeping of fish began in 1805. The first public aquarium in the world was opened in England in 1852, in the Zoological Gardens in Regrent's Park, London and since then keeping and breeding tropical fish has increased in popularity enormously. However, the market for ornamental fish for public aquaria in the world is less than 1 per cent at present. And over 99 per cent of the market for ornamental fish is still confined to hobbyists. The world trade of ornamental fish has been estimated to the tune of US $ 4.5 billion in 1995 and is moving further, with an annual growth rate of about 10 per cent per year.

Fish keeping means providing such conditions in the aquarium that the fish will live and fish remain in good condition for several months or years, depending on the life-span of the given species.

The primary principles of successful fish keeping are proper water conditions, composition, temperature, amount of oxygen and carbon dioxide in the water and correct feeding. Other important factors are lighting, planting and shelter in the aquarium, composition of the bottom substrate and suitable selection of fish. Fish keeping now a day is greatly facilitated by modern technology.

A basic rule is never to keep big and small fish, predatory and peaceful fish etc. together in a single aquarium. Do not skimp or space and always choose a larger rather than a smaller aquarium. The bigger the aquarium the less work there is and the greater the success of the aquarist. The aquarium should be located permanently in a spot where its fit in within the layout and furnishing of the room. The aquarist's successful in raising fish in captivity is due to great part to the fact that most freshwater and brackish water species are adaptable as to their environmental requirements.

Another decisive factor in raising fish is food. In nature fish are a link in a certain food chain, *i.e.* phytoplankton, zooplankton, small fish, predatory fish, man. It is impossible to provide the whole chain in the aquarium but we must know which parts in such chain the fish in our aquarium need.

According to the type of food they eat fish are divided into the following groups.

1. Carnivorous
2. Piscivorous or predatory fish that feed on animal food.
3. Herbivorous or vegetarian fish that feed only on plant food.
4. Omnivorous fish that feed on both animal and vegetable substances.

Depending upon type of fish, food should be provided for keeping healthy fish in aquarium.

Concept of Sustainable Development Applied to Fish Keeping

Natural sources of fish stocks are often subjected to inefficient use, often destructive, with irreversible effects in the long term and thus anti-economical. To ensure efficient use of the natural resources, the parameters for sustainable development must put in place. Sustainable development will achieve when all the environmental, social and economic priorities for mutual benefit. (Katia Oliver, 2001)

Present Status of Ornamental Fish Trade in India

Almost 70 per cent of the world trade in Ornamental fishes is in the hands of Asians. The principal players in this trade are Singapore, Hong Kong, Thailand and almost every country in South–East Asia. In the world there are 140 countries involved in aquarium fish trade.

India's share in this global aquarium fish trade is only 0.007 per cent. In terms of money it comes to Rs. 23 million or two crores and thirty lakhs. However, India has a domestic market of Rs 10 crores in ornamental fish trade can this is fast growing. M.P.E.D.A. (Marine Products Export Development Authority) has estimated that India's potential for trade in aquarium fish is 5 billion U. S. dollars which are earned by the export of Ornamental fishes. "Nabard" had visualized the rise in out trade up to 1 per cent of the global trade in five years.

At present, our centers of trade and even in the breeding of ornamental fishes are Kolkata, Mumbai and Chennai. West Bengal is in the forefront with a share of around 90 per cent of our export earnings. Around 7 to 10 thousand people are engaged as part time breeders of Ornamental fishes in West Bengal. The ornamental fish market of Kolkata is locally called "Hatibagan Haat" and is the largest in the country. Along with fish sale, it also sales aquarium accessories, fish feed, aquarium etc. It opens only of Sunday and the fish trade on single day runs in to 2–3 lakh rupees minimum. (Durve, V. S., 2005)

Prospects of Commercial Production in India

The export of ornamental fishes from the country at present is mainly confined to freshwater varieties and the export is limited to the fishes from north–eastern states (85 per cent) and few bred varieties of exotic species (16 per cent) from other parties. In spite of the availability of rich ornamental fish fauna in and around coral reef of Lakshadweep, Andaman and Nicobar Islands and Mandapam area, the country could not make any headway in the export of marine ornamental fishes so far due to non-availability of required infrastructure facilities. As the country possesses vast resources in term of natural water bodies and species diversity, we have a great potential to increase the level of export of ornamental fishes to about Rs. 110 crores every year.

The prospects of ornamental fish breeding and rearing along with the associated problems were discussed at a workshop organized by MPEDA at Chennai on 16th May 1997. The requirements as suggested at the workshop are as under:

1. Commercial ornamental fish production farms with necessary infrastructure facilities and with due support by technical experts are necessary to be set up for mass production of ornamental fishes.

2. Farmers with small scale production capacities need to be provided with necessary support facilities for enhancing their output.

3. It is necessary to equip the farms with fish pathology laboratories to over come possible disease outbreaks.

4. Health centers for quarantine certification at key location need to be established to meet the desired standards of the importers.

5. Ornamental fish culture sector needs to be provided with similar facilities and concessions as given to agriculture sector.

6. Information on ornamental fish breeding, accessories required for aquarium management, different management techniques including nutrition and disease; market situation; information on export potential and possibilities; transportation, packaging and freight systems; handling of fish from production to destination; climatic conditions of importing countries; organization of seminars/workshops, trade fairs and exhibitions in different parts of the world etc. may be made available to the farmers and exporters regularly through different publication sources.

7. Training programmes on different aspects of aquarium management for short and long durations depending on the clientele groups and requirements have to be organized at institutes/organizations having expertise for the purpose.

(Swain, Jena and Ayyappan, 2001)

Overview of Demand Situation

The number of species known as ornamental is estimated at about 1600 of which 750 are from freshwater and the remainder marine species, which only represent 10 to 20 per cent of total volume. Consequently the species that dominate the market are all from freshwater.

The main importers of ornamental fish are the United States, Japan and Europe. About 90 per cent of imports of ornamental fish into the USA are traditionally from Asia and 10 per cent from South America. The US imports from Indonesia and the Philippines are mainly composed of marine species.

Advantages of Ornamental Fish Culture in India

Advantages of ornamental fish culture in India is as under:

1. Both fresh and marine ornamental fish are available in India.
2. Being a tropical country ornamental fish culture is economically viable. Fishes can grow and attain maturity very fast.
3. Maintaining an ornamental fish farm (both hatchery and grow out) is very easy, particularly a freshwater ornamental farm.
4. Breeding can be done through out the year.
5. As most fishes breed naturally, the cost of hormone for breeding purpose can be avoided, unlike freshwater edible fishes.
6. Ornamental fishes have a thriving domestic and international market.
7. Ornamental fishes are generally more hardy; they can tolerate a wider range of fluctuations in the water quality parameters and less susceptibility to disease infection.
8. A lot of new opportunities exists for employment generation in the rural sector, particularly in the coastal areas where unemployment is a hindrance to the development process.
9. Unlike prawn farming, ornamental fish culture requires only a small area.
10. Ornamental fish farming is very profitable venture which requires less initial investment compared to prawn farming.
11. As the investment is less, the risk of production is also less.

Aquarium

Feng Shui: For Wealth and Health

Let us Brighten World by Me

Fish is generally regarded as one of the symbols of Wealth and Good Fortune. Feng-Shui means Wind and Water, these two are powerful forces of nature. By means of Aquarium we are able to keep Air, Water and Fishes in beautiful, sparkling manner in home. This is very simple and important way of increasing our growth and expansion. If you are arrange your aquarium in your house, office or factory in right direction, it brings stream of money to you.

If in your aquarium one or more fishes died suddenly, please do not be worry. Put new other same fishes in aquarium, because it is strongly believed that when this happens, it is no cause for alarm, they merely succeeded in warding some serious bad luck for the resident.

Best place for keeping aquarium is drawing room or living room. True direction for this is east, east-south and north. According to the Bagua all directions have their own merits. Never keep aquarium or fish bowl in bed-room, kitchen or toilet because this is believed harmful. Please never keep your aquarium in right side of your house main door, when you are going out side, because this make your brain disturb.

Normally for Feng-Shui fishes are kept in even number. Gold Fish and Arowana fish is referred as real Feng-Shui fish. Very beautiful Arowana fish is very expensive and fully grown can some time fetch several thousand rupees. Especially when their scales have noticeably transformed from the original silver colour in to either Gold or Pinkish red. It is believed that this is a clear indication that the millions are coming.

If to keep live aquarium is not possible, please keep beautiful photograph of fish aquarium. This will also give you some good-luck.

Shower your Love

Win our Love

Feng Shui–for Aquarium Keeping

Feng-Shui Aquarium

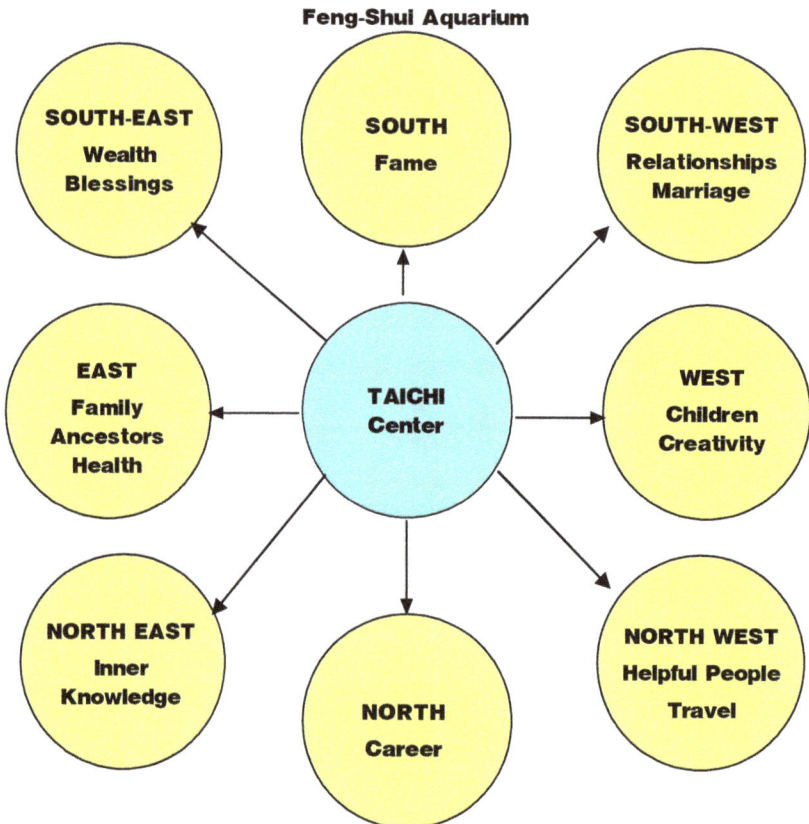

Water is only beneficial if it is kept clean, clear and sparkling, as this enhances C. H. I.

Shape of Aquarium is not so important. Healthy Red and Black fish also enhance C. H. I.

How to Develop Ornamental Fish Farming

Although ornamental fish culture is an age old practice in India, till now it is not a popular enterprise. So, more exhibitions, training programmes, seminars and symposiums are required in order to popularize ornamental fish farming. For ornamental fish farming following suggestions may be considered.

☆ To the resource poor farmers engaged in ornamental fish culture, the concerned government should provide financial assistance in terms of long term loans with adequate subsidies.

☆ Many of the fisher folks do not know the value of ornamental fishes and they use them as food. So, a general awareness should be created among the fishing community about these much valued organisms.

☆ As ornamental fishes are of great attraction to everybody, marine/freshwater aquaria can be set up in tourist area. They will be good source of revenue through a nominal entrance fee.

☆ Like they do in the case of prawn farming, MPEDA and other export promotion agencies should come forward to offer technical and financial assistance to the farmers who enter this virgin field.

☆ Marine Ornamental fishes fetch good prices compared to freshwater variety. But the major problem is that breeding through artificial propagation has not been standardized on a commercial scale.

☆ An ornamental fishes are to be transported live, greater importance should be given to conditioning prior to packing in order to minimize the risk of mortality due to stress.

☆ As the fisher folks are not well versed in catching marine ornamental fishes. They often unwittingly destroy coral reefs, the real habitat of these species. Therefore, suitable gear should be provided and scientific catching methods are to be developed to conserve the natural resources.

☆ In ornamental fish culture, the maximum mortality is found within 3–4 days after hatching when the yolk sac is consumed fully by the young ones and the mouth is to be formed for taking natural food. If natural food is not suitable to them, they die immediately. Therefore, live food culture is a must like in the case of prawn culture in order to avoid premature death of the fishes.

Chapter 2
Fabrication and Arrangement of Aquarium

Fabrication of All Glass Aquarium Tanks

Before fabrication of all glass aquariums, one must first decide about the size of the tank. Different thickness of glass is required for different sizes of tank. There should be a strict relationship among length, depth and height to suit the principle of water pressure which exercises pressure against the bottom and the four walls, the front, the back and the two sides. Length of aquarium is the deciding factor, it must be twice the depth while height equal to the depth. Table 2.1 gives general details of size of aquarium and thickness of glass required.

Table 2.1: Showing Size of Panes of Glass Required, Glass Thickness, Water Capacity and Number of Fish can be Kept

Tank Dimensions (Inches)			Tank Dimensions (cms)			Glass Thickness	Surface Area	Water Capacity (Liter)	Weight of Water (kg)	Maximum Fish Capacity (Body size) cm
Length	Width	Depth	Length	Width	Depth					
18	10	10	45	25	25	4	1125 cm²	27.3	27	38
18	15	12	60	38	30	6	1350 cm²	45	45	45
24	12	12	60	30	30	6	1800 cm²	54.6	54	60
24	15	12	60	38	30	8	1800 cm²	68	68	60
36	15	12	90	38	30	10	2280 cm²	104	104	90
48	15	12	120	38	30	12	1300 cm²	136	136	120

The glass sheet should be got cut into correct sizes of the panes for the tank size selected. All edges should be evened by filing with wet fine grain sand paper or sand stone. Clear all edges. Final view of all glass aquarium will look like Figure 2.1 (a), (b), and (c).

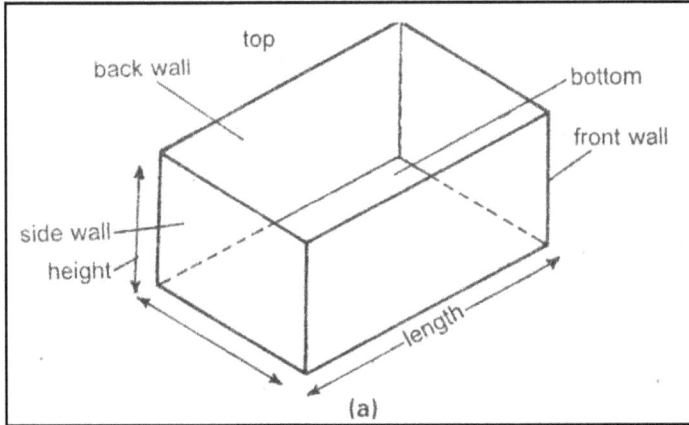

Figure 2.1(a): Front View

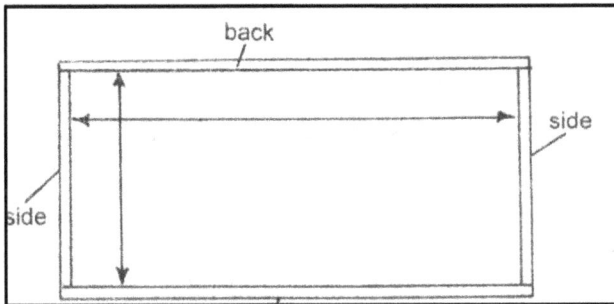

Figure 2.1(b): Top View

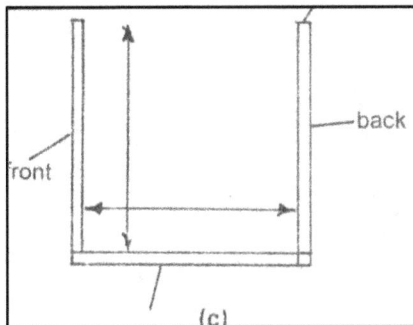

Figure 2.1(c): Side View

Figure 2.1: Fabrication of Aquarium Tank

Place the base glass on a flat surface and stick pieces of adhesive tapes on all the four sides (Figure 2.2a). Make a glue line of silicon adhesive using the nozzle of the gum on the edge that will hold the back glass pane (Figure 2.2b). Join the back glass pane to the base on this glue line, stick the flying adhesive tapes on to it and provide weight from outside to give support to it. Stick adhesive tape pieces to the sides of the glass pane. Likewise repeat it on the other side to join the second side glass pane. Join the front glass pane in the last. Silicon glue dries up within minutes, so it is essential that any surplus glue on the inside be wiped out immediately using a solution of washing up liquid in water. Extra care may be taken to seal corners to get water tight aquarium tank.

Leave the cemented tank exposed to air at least for 2–3 days so that the bond is perfect. To check any possible leakage fill it with water and allow it to stand over night. Repeat it three times. Now the water tight tank is ready. To strength glass corners, plastic or metal corners may be glued using silicon glue (Figure 2.2c).

Figure 2.2: Fabrication of Tank

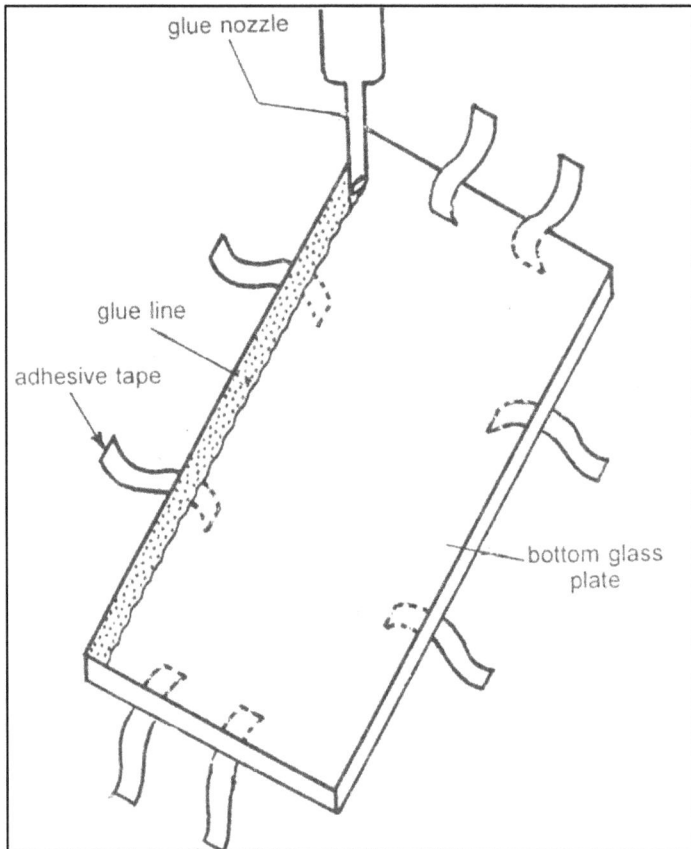

Figure 2.2(a): Making a Glue Line

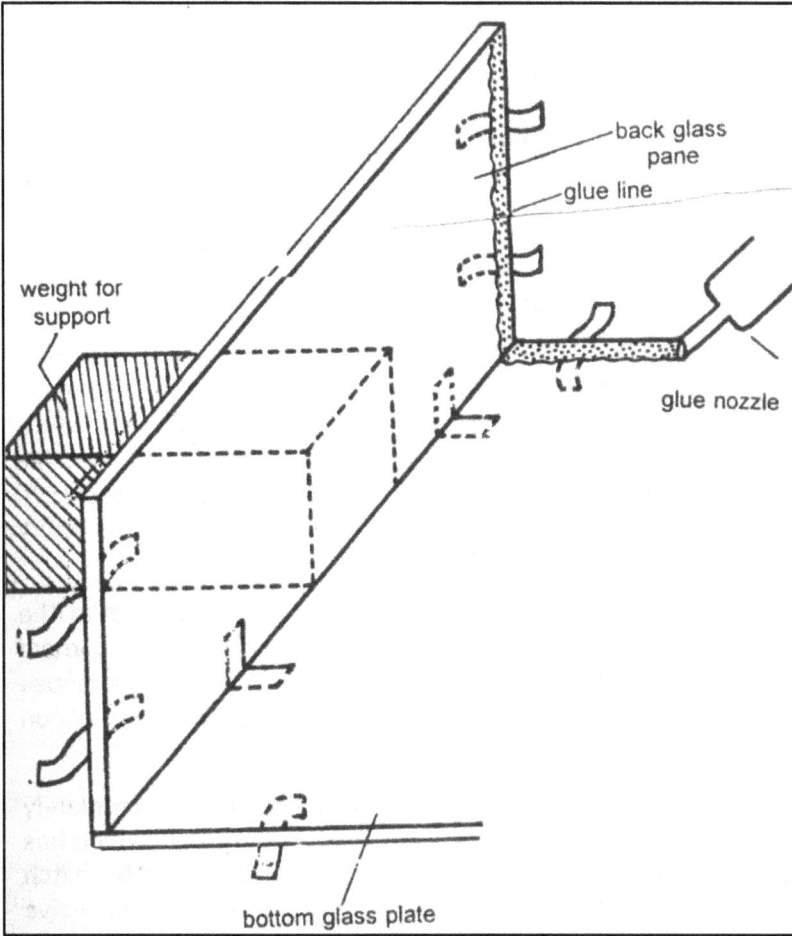

Figure 2.2(b): Cementing the Back Wall at the Base

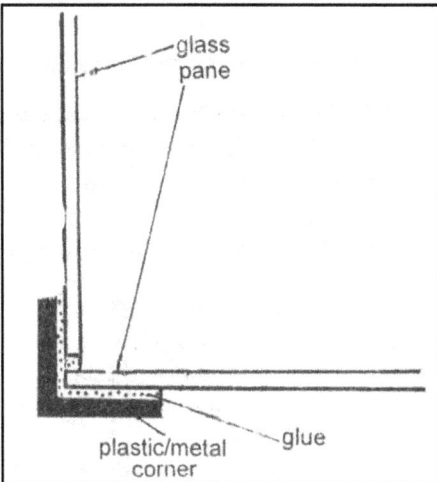

Figure 2.2(c): Cementing Corner

Fabrication of Framed Aquaria

Such aquaria are made of aluminium or iron frames. The width of aluminium frame is generally 3/4–1 inch. Large aquaria (size 36″ × 15″ × 15″ or more) are made of iron angle frame. For such aquarium tanks glass should also be thick *i.e.* more than 6 mm.

1. For making the framed aquarium of 24′ × 12′ × 12′ a similar sized frame is fabricated using aluminium angle riveted with aluminium rivets. The holes for rivets are made about 1/4″ from the edge of angle.

2. For this aquarium a glass of 6 mm thickness is used. The bottom glass pen should have a size of 61 × 30.5 cm, front and back panes of the size 61 × 29.9 c. and both side panes of the size 28.3 × 29.9 cm. First of all the edges of all glass sheet should be grounded carefully using a glass grinding stone. The bottom glass is first placed in the frame followed by the front glass sheet. Then both these glass sheets are stuck together from the corner.

3. To fix the glass in the frame and together, aquarium cement or battery compound (bitumen) is used. The compound should be heated and carefully poured in the glass corner using wooden or metallic spoon.

4. The back and two side panes are then fixed in the frame using the same procedure.

5. Next day, after proper cooling and setting of cement, the glass tank may be checked for leakage by filling it with water. The leakage can be stopped if the tank is added with some fine sand. The sand particles will clog the leakage point in few days. (L.L. Sharma, 2005)

Importance of Cover

No aquarium is completed without a glass cover or non toxic hood. It should fit loosely on top of the aquarium to allow circulation of air. Apart from checking fish to jumping out, covers on aquarium slow up loss of water by evaporation and stops a lot of dust and other bodies from getting in. It also provides support for lighting system. Usually a wide slit with mesh cover is left on the sides of aquarium hood for aeration.

Top covers of aquarium gives attractive look to aquarium. At present in market, tops are fabricated with plywood and fiberglass with lighting facilities. Some aquarists prepare top with PVC strips, Bamboo strips, Glass sheets, PVC electric pipes and strips.

Setting Up the Aquarium

Setting up aquarium should be an enjoyable experience. A little "forward Planning" makes all the difference. You should bear in mind that the aim is not only to provide the fishes with their required conditions but that the aquarium can also be easily serviced. If you are a new comer to fish keeping make sure, before you start, that you clearly understand how the various pieces of equipment work and how they fit together.

Before actual setting up the aquarium following points are to be checked.

Selection of Site

☆ A firm, level base must be chosen.

☆ Make sure the weight is distributed evenly, preferable sharing the load over floor joints.

☆ A window location is not suitable; day light is uncontrollable and too much direct sun will overheat the tank and produce extra algal growth.

☆ The place chosen must be seen from all the angles inside the room.

☆ The height of place where the aquarium is must be in the same level of our heads, when we are comfortably sat enjoying our aquarium

☆ We must avoid that the aquarium receives direct sun light.

☆ It can provoke an over heating of the aquarium and provoke a truly "boom" of algae.

☆ The aquarium can turn completely green what becomes unpleasant and unaesthetic value.

☆ Do not place an aquarium above or near a heat source, heating vent or stove.

☆ Avoid high traffic areas as the constant movement and noise will tend to stress the fish.

☆ Convenient and plentiful electrical outlets are required and a water source, inclusive of drain, within fifty feet of the aquarium is an added convenience.

Requirements

1. Aquarium with cover glass and hood
2. Light
3. Heater/with thermostat
4. Filtration system (mechanical, chemical or biological)
5. Air pump, air line and air stones
6. Gravel (and any additional base covering material like sand)
7. Plants
8. Suitably treated water–Not chlorinated water straight from the tap.
9. Scissors, pliers, screw driver and electric plug.
10. Background Posters
11. Decorations
12. Net

13. Gravel vacuum

14. Dechlorinator

15. Algae scraper

16. Test kits

17. Food

18. Thermometer

19. Freshwater aquarium salt

20. Aquarium book

Preparations for Aquarium Setting

Before setting up, test the tank for leak (out side) and wash gravel and rocks. If you intend making up rocky structures as tank decorations, these can be made in advance.

☆ Plants should be rinsed and inspected for snails' eggs and other unwanted passengers.

☆ The lights and pump are connected to the switched terminals. Mount the cable tidy on the end of the tank nearest the power supply socket.

☆ All glass aquariums should have an absorbent cushion beneath them to iron out any unevenness in the surface on which the tank will stand. A sheet of expanded polystyrene cut to the dimensions of the tank base is ideal.

☆ If biological filtration is to be used, the filter plate must be put into the tank before the base covering gravel is added. Make sure that the filter plate is well bedded down and covers the whole base; failure to do this will result in water finding its way around the plate instead of going through the gravel, and if the plate does not cover the whole base some areas will become stagnant and anaerobic (lacking oxygen). Fit the vertical tube to the filter plate now otherwise gravel will get into the hole later.

☆ Put large rocks directly on the filter plate or the tank base. In this way they will be supported by the rest of the gravel and be less likely to topple.

☆ Spread washed gravel over the filter plate (if fitted) or directly on the tank base to a depth of 2–3 cm.

☆ Fit the heater/thermostate combined unit to the rear wall of the tanks.

☆ Hide the hardware and pipe work with strategically placed rock work but do not obstruct water flow to the filter's inlet tube. View the set-up frequently through the front glass to check satisfactory artistic progress.

☆ It is now time to fill the tank with water. This should be done gently to avoid disturbing the arranged gravel using a hose let water first flow into a saucer or small jug placed on the gravel. The overflow will fill the tank without disturbance. When the tank has only few centimeters left to fill, stop the water supply.

☆ Now is the best moment to plant tank, the plants will take up their natural positions in the water and you will be able to judge the effect instantly.

☆ Rooted plants should be placed in holes in the gravel. The junction between plant root and stem should be just clear of the gravel surface.

☆ Add bushy species to fill the corners.

☆ Do not be too sparing with plants. It is quite common to have 50 plants per square foot.

☆ The aquarium lights should come on, bubbles emerge from air stones.

☆ The water flow from under gravel filters need not be too excessive a steady flow is all that is required.

☆ Now put fish which is previously washed with clean water of proper nature and of proper size to suit the size of aquarium.

☆ Figures 2.3–2.6 give some arrangement patterns. However, arrangement in aquarium is individual skill.

Stocking Capacity of an Aquarium

☆ How many fish you can keep in a particular sized aquarium.

☆ The most common formula for the amount of fish you can safely keep are based on the number of liters of water in the aquarium.

☆ Keep 1 inch of fish for each liter of water.

Adding Equipment and Non-Living Decorations

☆ Position the equipment towards the back of the aquarium.

☆ Heaters, thermostats and filters should be attached to the glass using the suction cups provided.

☆ The wires from the electrical appliances (heater, thermostat, air pump, light etc.) can be connected to a cable tidy to reduce the number of unsightly wires around the back of the tank and make controlling the system much easier.

☆ Air pumps should be positioned above the level of the water. Fit the heater/thermostat combined unit to the rear wall of the tanks.

☆ Air valves, which control the airflow from the air pump top air stones and air operated filters, are best ganged together in a block in one convenient place.

☆ Connect one end of a length of air line to an air stone and its other end to an outlet on the block.

☆ Connect another air line from air lift tube of the internal/external filter box (or under gravel filter) to another outlet.

Placing Heater

☆ Heater is necessary during winter season because most tropical fish need warm water.

☆ This means they need a heater to not only keep the water warm, but to keep the temperature consistent as well.

☆ There are many heaters available, and they come in a variety of sizes to match the size of your tank.

Install Background

☆ A background can greatly improve the overall appearance of an aquarium.

☆ It hides the various cords and airlines that are usually dangling behind the tank, and it can also provide an attractive color or scene to view your fishes against.

Figure 2.3: Different Arrangements in Aquarium

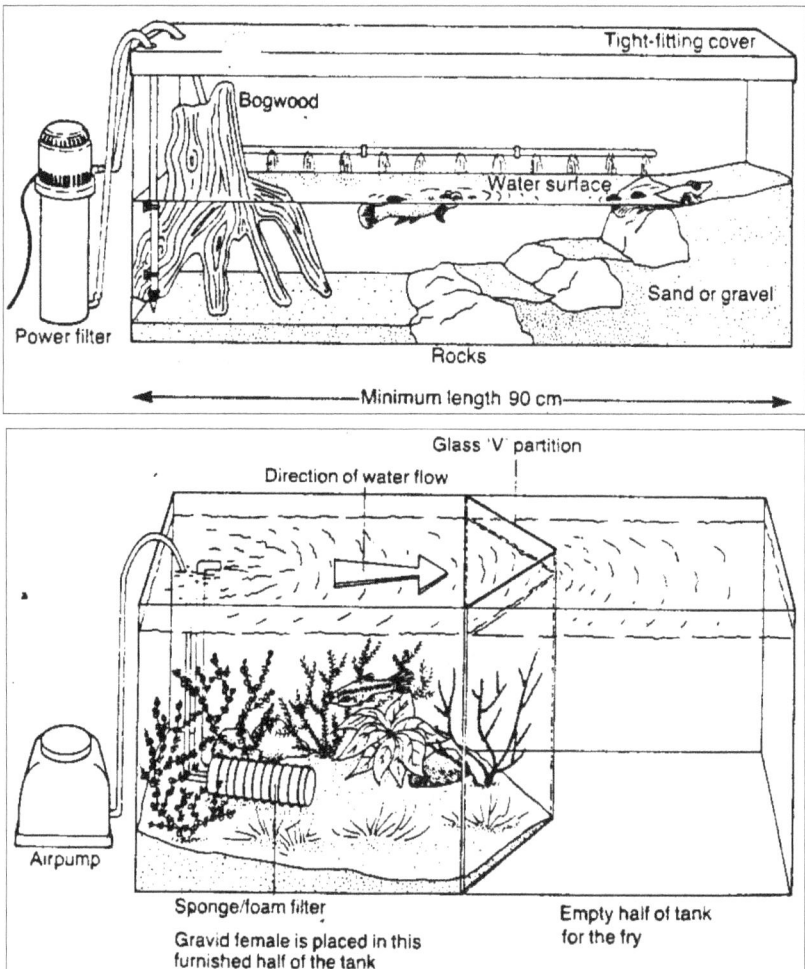

Indigenous and Highly Effective Breeding Tank

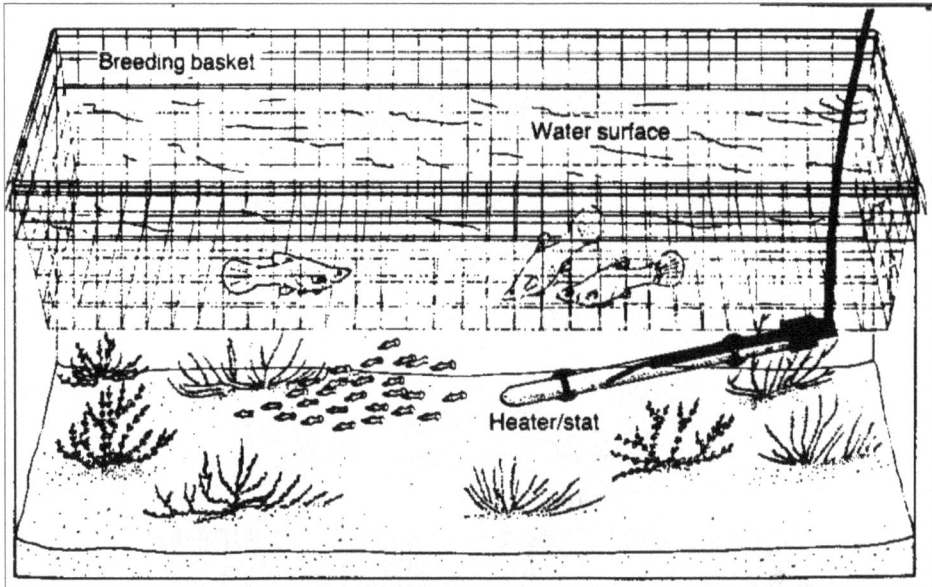

Breeding Basket provides all the Advantages of a Breeding Trap

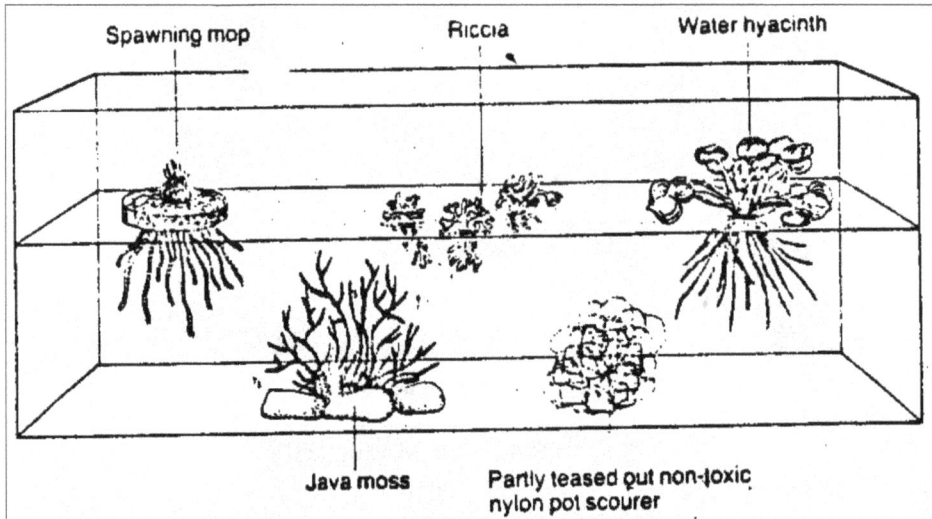

☆ Now is the perfect time to install an aquarium background, before the tank is full of water and up against a wall.

Different Shapes of Aquarium

Most shops stock a range of standard aquaria of different shapes *e.g.* simple gold fish bowl (Figure 2.7), normal shape with cover (Figure 2.9), triangular (Figure 2.11),

Arrangements in Aquarium

Figure 2.4: Arranging Corals

Figure 2.5: Arranging Plants

Figure 2.6: Arranging Gravels

Different Shapes of Aquarium

Figure 2.7: Round Shape

Figure 2.8: Octagonal

Figure 2.9: Normal Shape with Cover

Figure 2.10: Rectangular with Square Top

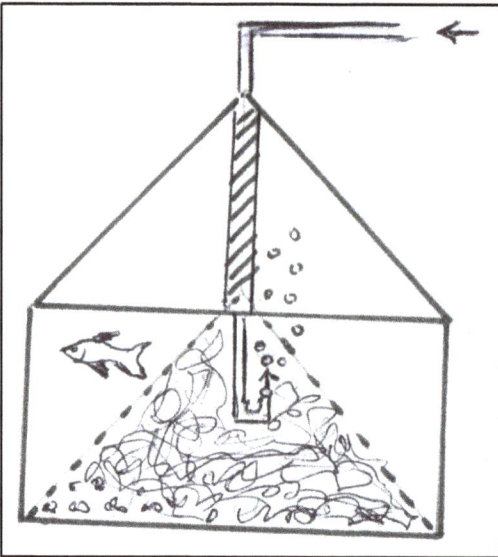

Figure 2.11: Triangle Aquarium

Figure 2.12: Special Shape

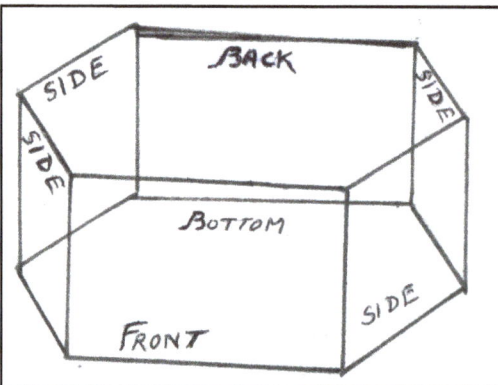

Figure 2.13: Octagonal Aquarium

Figure 2.14

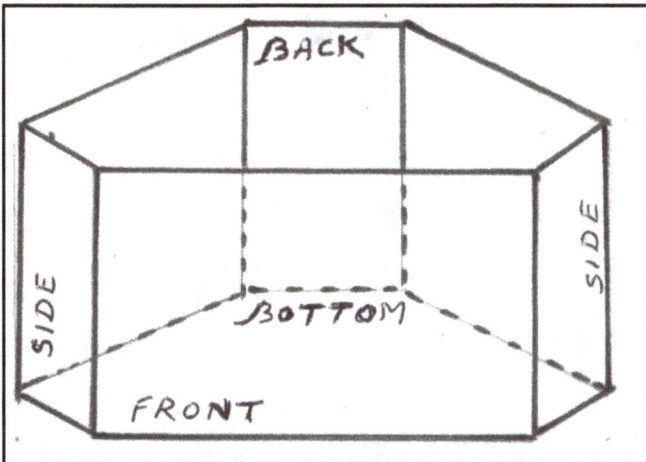

Figure 2.15: T.V. Type Aquarium

aquarium with square top (Figure 2.10), other special shapes are special shape (Figure 2.12), octagonal (Figures 2.8 and 2.13), aquarium in cupboard (Figure 2.14), hexagonal-TV type (Figure 2.15) or aquarium as a center table. Preference should be given to a rectangular aquarium since its proportionate size gives it an elegant look and firmness for handling too. Some of the standard sizes of aquarium are given in Table 2.1.

Adding the Gravel

☆ A 2-3 inch layer of gravel should be added.

☆ Slope the gravel towards the front of the tank so that any debris accumulates here and is easier to remove.

☆ Different types of Gravels are shown in Figure 2.16.

Installation of Under Gravel Filter

☆ The best all around filter is the under gravel filter.

Figure 2.16: Different Types of Gravels

☆ It is essentially a piece of plastic with small holes or slots in it.

☆ In one corner is a large round hole where a plastic "uplift" tube fits.

☆ An air pump and tubes are connected to the uplift tube.

Costing Calculations for Preparing Aquarium (Tentative)

Size of Aquarium	2.5' × 1.5' × 1'		
			Total
Size of glass	2.5' × 1.5'	2 Nos	7.5'
	2.5' × 1.0'	1 No	2.5'
	1.5' × 1.0'	2 Nos.	3.0'
	Total		13.0'

Cost of glass 5 mm glass @ Rs 35/-	Rs 455/-
Cost of glass 6mm glass @ Rs. 40/-	Rs.520/-
Cost of silica jell tube	Rs, 75 to 100/-
Cost of Biological filter, Marble, aeration stone.	
Tube, connections Toys etc.	Rs. 120/-
Aerator	Rs. 80/-
Fabrication charge	Rs. 40/-

Chapter 3
Aquarium Accessories

Introduction

All the aquatic life in an aquarium is its software whereas all the accessories which are used to maintain the water quality and all the aquatic life in an aquarium is its hardware part. For proper housing of the software, hardware has to function properly and efficiently. Various accessories are installed in an aquarium as hardware. These accessories could be further classified as essential and decorative.

Essential Accessories

The accessories under this category are highly essential. These will help to maintain the water quality of aquarium and reduce your trouble of aquarium maintenance and also keep your fish healthy and alive for a longer period.

(A) Filters

Fish is constantly discharging its urinary and fecal waste in aquarium water, uneaten food also starts putrifying. All these increases ammonia load in water, which becomes toxic to fish. To control this either one has to constantly change water or use water filters. The regular change of water is quite cumbersome and also stressful to fish. Thus, the water filters are used. These filters provide substrate to various ammonifying and nitrifying bacteria and also absorb the remove the suspended particles in water (Atul Kumar Jain, 2005).

Different Types of Filters Used in Aquarium

1. Under Gravel Filter or Biological Filter

Fishes and invertebrates excrete ammonia as a waste product. This is added to the ammonia produced by bacteria working on other waste material in the aquarium, such as dead plants and uneaten foods. Ammonia is toxic to fishes and invertebrates and if it is not removed, or converted in to other less harmful substances, then aquatic life will soon perish.

Aerobic (Oxygen loving) bacteria of the genus *Nitosomonas*, convert ammonia to nitrite, a slightly less toxic substance but one that is still dangerous to fishes. A second group of bacteria, belonging to the genus *Nitrobacter*, transform the nitrite to nitrate, a much safer substance, but one that can still cause problems it is building up in the aquarium. If further anaerobic bacteria are allowed to get to work, then the nitrate can be converted back to atmospheric nitrogen and vented from the aquarium.

In the basic set up, it is used to provide the conditions required by the first group of aerobic bacteria *i.e.* those that thrive on oxygen. It is therefore, necessary to provide under gravel form of filtration. An under gravel filter is a perforated plate covered by a 7 to 10 cm layer of suitable coral gravel and coral sand. In a traditional "down flow" filtration set-up, polluted, but oxygenated, water is drawn downwards through the substrate, where colonies of nitrifying bacteria living on the surface of the substrate particles break down nitrogenous waste in two stages.

Under Gravel (UG) filters are perforated plates of plastic [Figure 3.1(a): Principle and Figure 3.1(b): Actual]. These are available in various sizes as per the dimension of aquarium bottom and covered with gravel. At one corner of the plate a pipe is attached through which an air supply tube also passed to the aquarium bottom, Terminal end of the air tube is provided with a diffuser stone. Water of the aquarium passes through gravel and accumulates below UG plates. During this process toxic gases are removed from the water due to action of bacterial colony which develops in gravel bed. The other solid waste also settles in gravel and slowly acted upon by

Different Types of Aquarium Filter

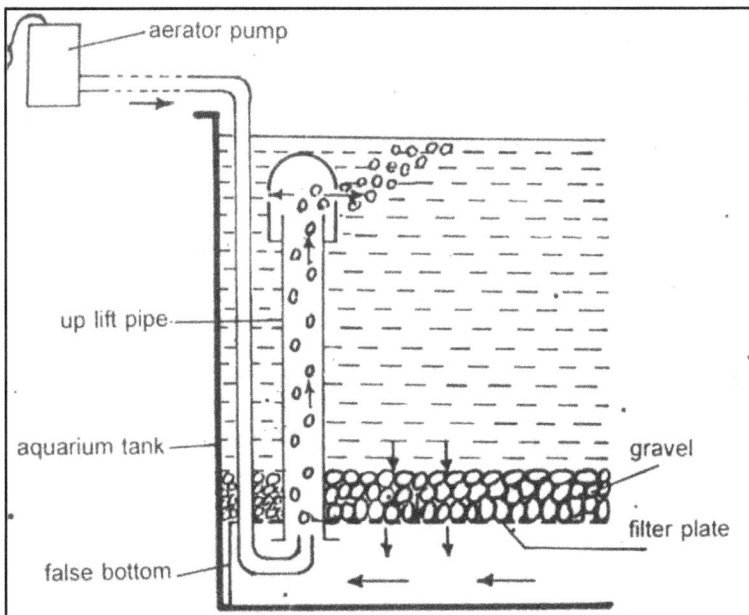

Figure 3.1(a): Under Gravel Filter (Principle)

Figure 3.1(b): Under Gravel Filter (Actual)

bacteria. When air is pumped to the aquarium bottom, the filtered water accumulated below the UG plates get lifted up and released back in aquarium. This recirculation process maintains the cleanliness of water for a longer period.

2. Air Lift Box Filter

It is a small plastic box (Figure 3.2) filled with filter media (Charcoal, foam, small pebbles etc.) The working principle is same as of UG filters. However, these are easy to install and clean as it is independent unit.

3. Sponge Filter (Figure 3.3)

It is modification of sub gravel biological filter in this case; gravel bed of the aquarium is replaced by sponge/foam bed through which aquarium water is filtered before being air lifted for returning it to the aquarium.

4. Mechanical Filter

Mechanical filtration serves the simple function of removing visible

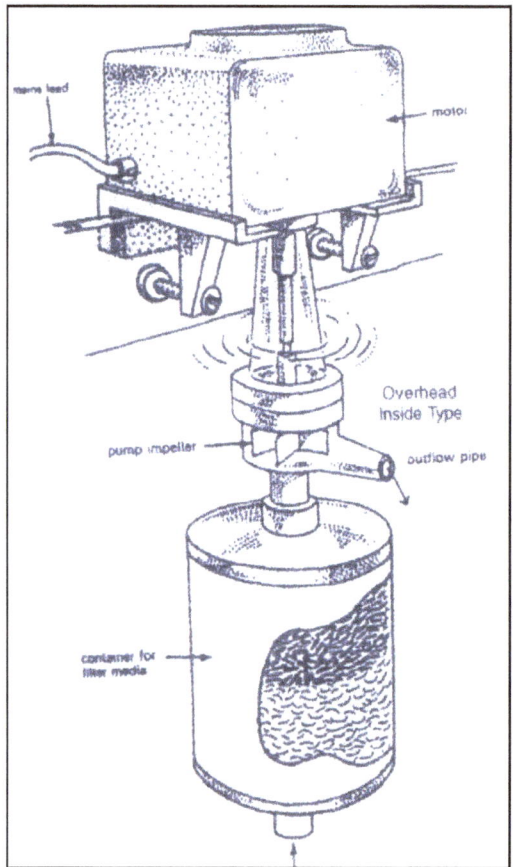

Figure 3.2: Airlift Filter

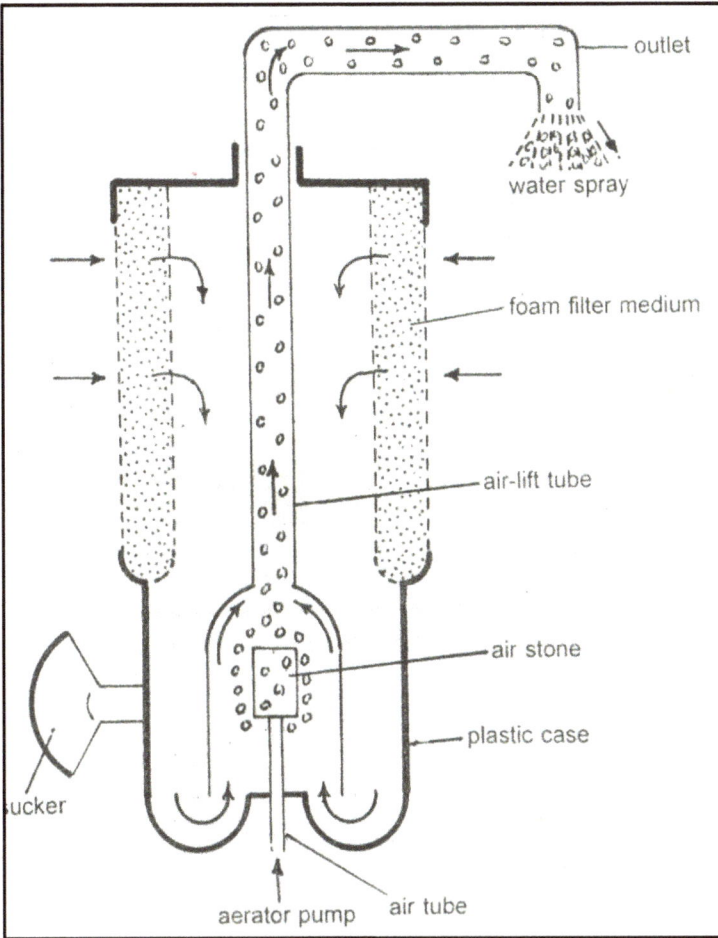

Figure 3.3(a): Mechanical Filter (Sponge type) (Principle)

Figure 3.3(b): Sponge Filter (Actual)

particulate matter from water and delivering it to a point where it is easily removed and/or where it can be broken down in to less toxic end products.

The most commonly used mechanical filter medium is some form of floss made from spun nylon, or some other man made fiber. Many fish keeper use plastic pot–scourers preformed foam pads in fitted in to the filter body not only perform a simple filtering function but, of not renewed on a frequent basis, also acts as biological filter once bacteria have colonized the foam (Figure 3.4).

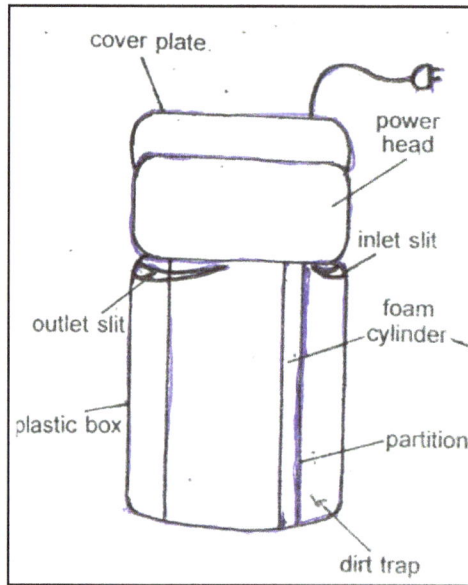

Figure 3.4: Mechanical Head Filter (Submersible Type)

5. Chemical Filter (Figure 3.5)

The most common chemical medium used in canister fibers is activated carbon. The carbon is usually sandwiched between two layers of floss in the filter body to prevent it from being drawn in to the aquarium. Alternatively, the charcoal can be contained in a nylon bag which serves the same purpose. This acts to remove ammonia, nitrates, phosphates or other pollutants. The carbon will absorb chemicals. It is suggested to renew every two months on average. If this is not done, the absorbed chemicals are likely to be released back in to the aquarium, thus negating all the beneficial works of this filter. In chemical filter different filter media are used. Type of media and its activity are as under.

Sl.No.	Media	Item Removed
1.	Activated charcoal	CO_2, pH and hardness
2.	Limestone chips	pH and hardness
3.	Coral sand	pH and hardness
4.	Resin (Zeolite)	Hardness and Ammonia

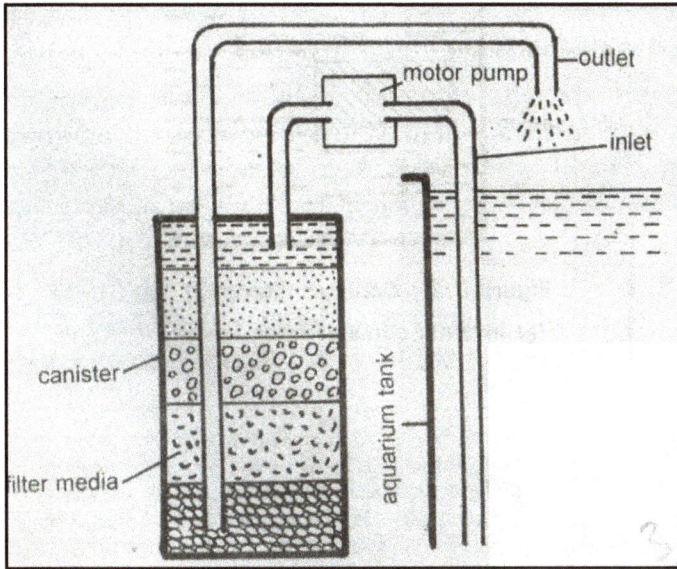

Figure 3.5: Chemical Filter

6. U. V. Sterilizer

U.V. light has long been known to have a strong sterilizing effect and U.V. penetrating units have many industrial uses as bactericides. Such units which pass pre-cleaned water from a power filter close by an enclosed U.V. light are now widely available to the fish keeper. The light they emit can kill algal spores and bacteria and may affect some small parasitic organisms.

The water of aquarium is allowed to pass through a chamber where it is exposed to U. V. radiation. Most of the micro organisms are destroyed, however, as the time of exposure to short some bacteria is not killed including the pathogenic forms. Thus the risk for bacterial diseases remains if proper cleaning and operation is not done.

7. Canister Filter (Power Head Filter)

It is a self-contained internal type unit comprising of a container, provided with an aerator. It operators on the principle that aquarium water is siphoned into filter unit where it is allowed to slowly pass through one or more filter or being pumped up or air-lifted and returned to the aquarium [Figures 3.6(a), 3.6(b) and 3.7].

It may be kept inside the aquarium. The filter media used include nylon floss, glass wool, foam, activated charcoal, resin etc. It may be used for mechanical, biological and/or chemical filtration.

Figure 3.6(a): Canister Power Air Lift Type (Actual)

It can be used as external filter also. The design is given in Figure 3.8.

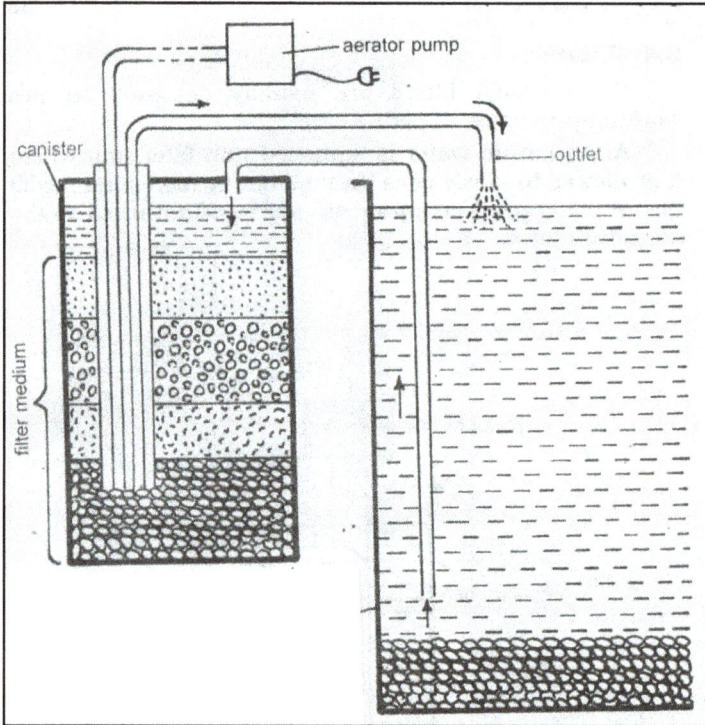

Figure 3.6(b): Canister Power Air Lift Type (Principle)

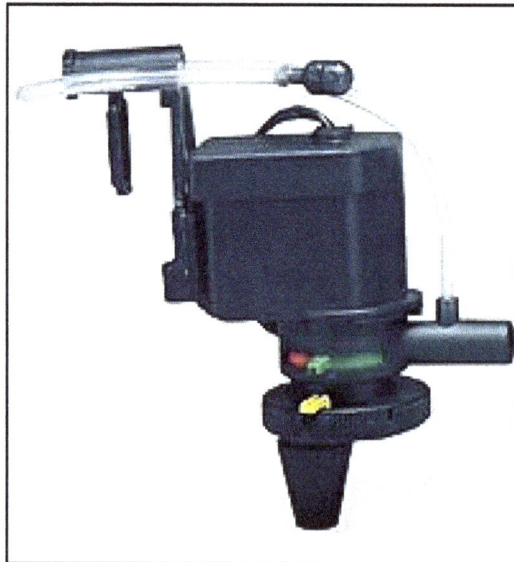

Figure 3.7: Power Head Filter

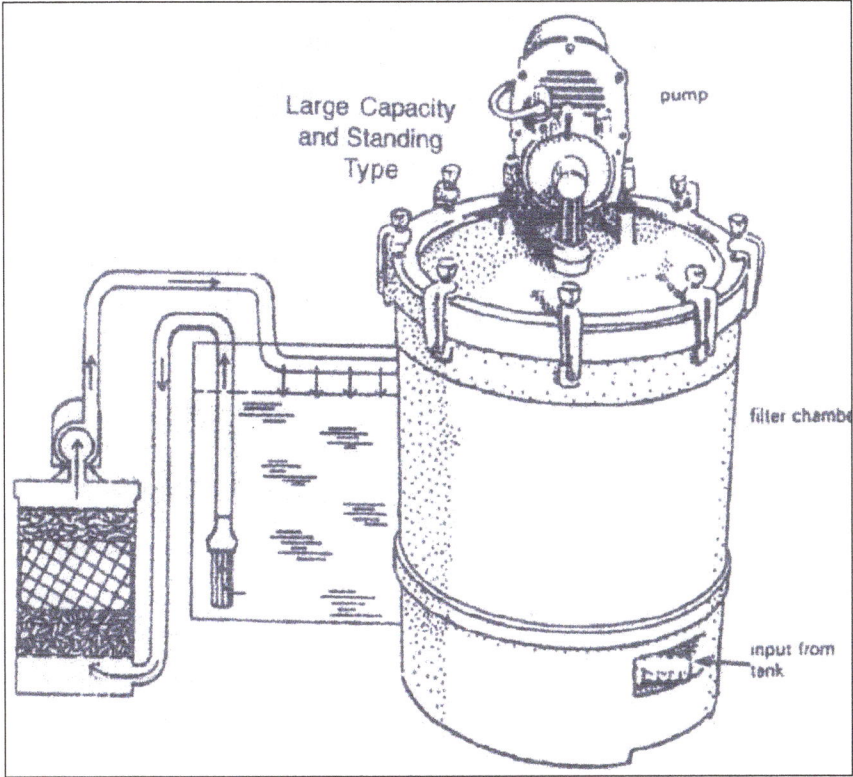

Figure 3.8: External Filter

Photographs of Some Filter

Figure 3.8(a): Fluidized Bed Filters

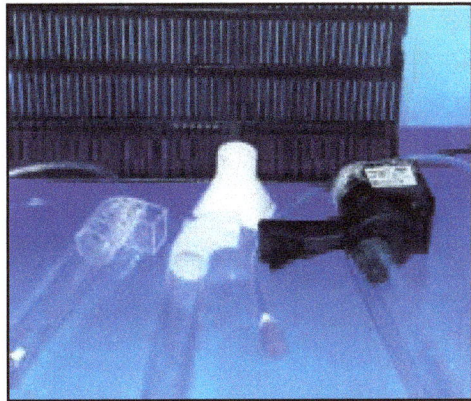

Figure 3.8(b): Under-gravel Filters (UGF)

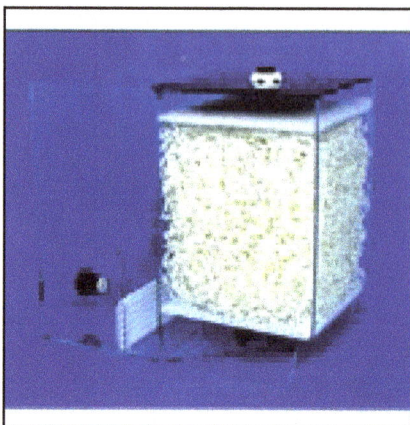

Figure 3.8(c): Box Filter (Simple variety)

(B) Different Type of Other Accessories Used in Aquarium

1. Air Line Accessories (Figures 3.9 and 3.10)

> (*i*) Clear flexible PVC tube
>
> (*ii*) 3 or 4 way connectors to distribute air to 2 or 3 routes
>
> (*iii*) Screw valves to control flow of air along one route.

2. Air Stone (Figure 3.11)

It is perforated cube which diffuses air in the aquarium water.

3. Hand Net (Figure 3.12)

Hand net is indispensable. It is used for introducing fish into the aquarium tank or for shifting fish from one aquarium to another.

Different Aquarium Accessories

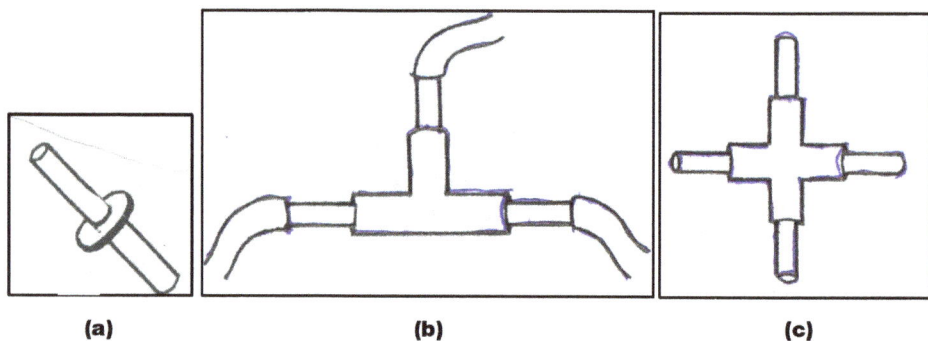

(a)   (b)   (c)

Figure 3.9: 2-3-4 Way Connectors

Figure 3.10: Screw Valve

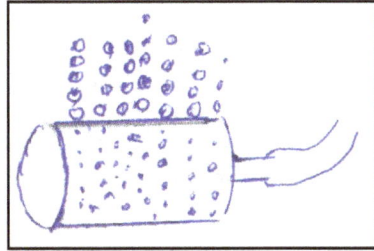

Figure 3.11: Air Stone

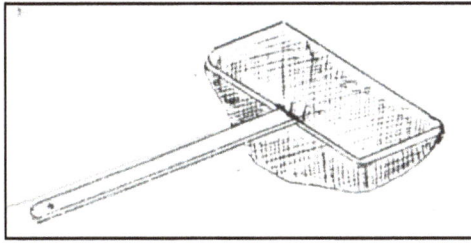

Figure 3.12: Hand Net

4. *Magnetic Scraper (Figure 3.13)*

It is used to scrape off algae from the glass surface.

5. *Worm Feeder (Figure 3.14)*

It is a very useful device for feeding worms to the fish. It consists of plastic basket perforated with slits. *Tubifex* worms or *Chronomus* larvae may be placed in the basket

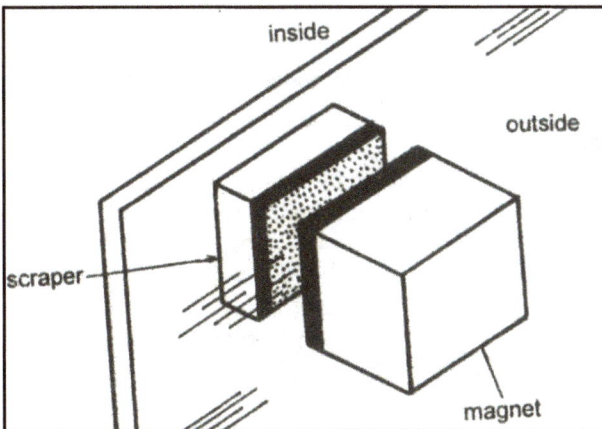

Figure 3.13: Magnetic Scraper

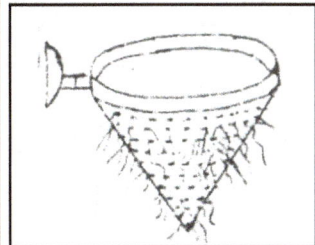

Figure 3.14: Worm Feeder

which is kept submerged in water, at a height near the surface. As the worm prop out of the slit they are eaten by fish (Srivastava, 2002).

6. Aquarium Gravel

Aquarium gravel is the most common bed material. It not only provides surface for aqua scarping but also provides rooting surface for the plants as well as serve as a good habitat for the helpful microbes. A number of materials such as marble chips, semi precious stones, cores sand or pieces of plastic materials can be used for this purpose. Broken shell can also be used. For aquarium of 24 × 12 ×12 inches about 5 kg of gravel is required. A grain size of 3–6 mm is suitable for most uses. It needs to be washed properly prior to use.

7. Silicon Gun (Figure 3.15)

Silicon adhesive is currently used in fabricating an all–glass aquarium.

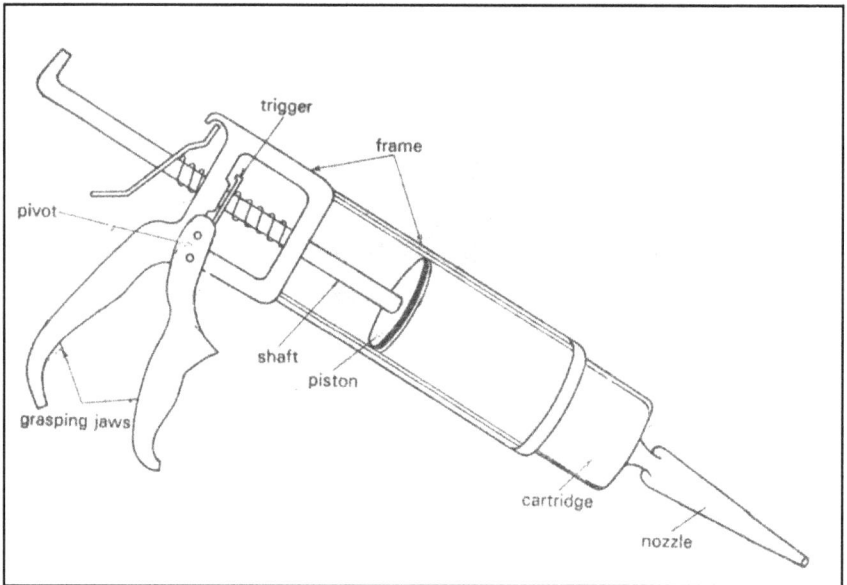

Figure 3.15: Silicon Gun with Cartridge

8. Submersible Heater (Figure 3.16)

Most tropical fishes require water heating during winter season. An immersion heater consisted in glass tube and attached with a submersible thermostat is used for

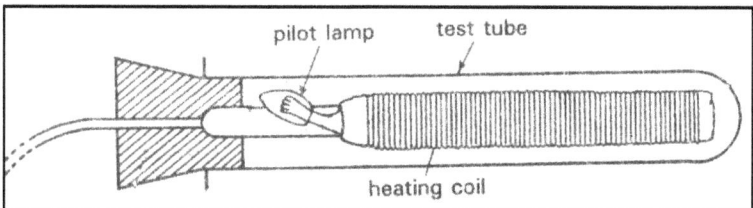

Figure 3.16: Submersible Heater

heating the aquarium tank water. The thermostat automatically cutoff the electrical supply of the heater after achieving a required temperature.

9. Thermostat (Figure 3.17)

It is an instrument to maintain tank water to desired temperature.

10. Aerator (Figure 3.18)

It is an instrument which supplies air to aquarium water tank to maintain oxygen level in the tank.

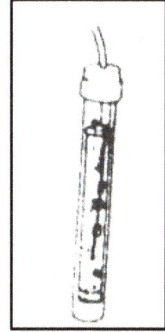

**Figure 3.17:
Internal
Thermostat**

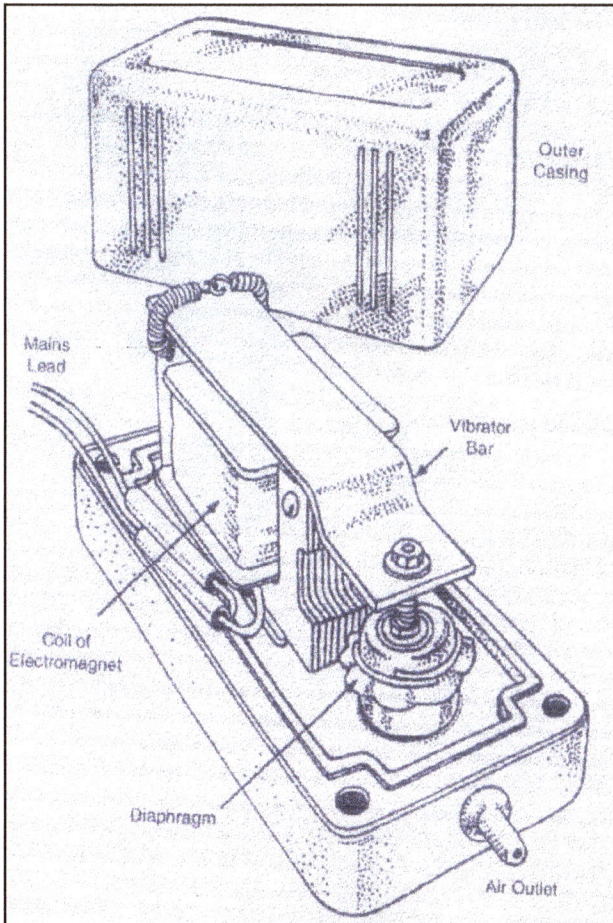

Outer
Casing

Mains
Lead

Vibrator
Bar

Coil of
Electromagnet

Diaphragm

Air Outlet

Fig: 3.18 Aerator

11. Toys (Figure 3.19)

Various kinds of decorative toys either fixed or working through air supply are available in market and fascinate a new aquarist.

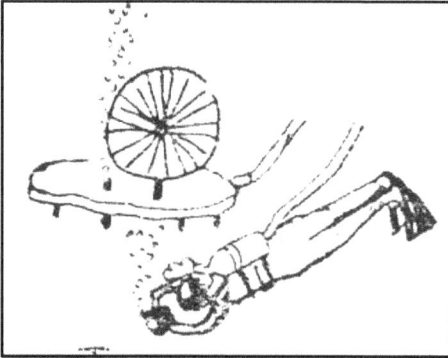

Figure 3.19: Toys

Figure 3.20: Artificial Plastic Plant

12. Plastic Plants (Figure 3.20)

Various types of plastic plants are available in market. One can use when natural plants are not available. Natural plants can decay while plastic plants are permanents.

Chapter 4
Aquarium Plants

Introduction

Aquatic plants can offer an ideal place for fishes to refuge and many fishes prefer aquatic plants and parts suitable for egg laying and building nests. Soft aquatic plants are useful as direct food for herbivorous fishes.

They can be divided in two parts:

1. Freshwater and
2. Marine water plants.

Freshwater Plants

Freshwater plants, which are ornamental and submerged, are widely used in aquarium tank. They take up the nitrogenous material from water and increase oxygen level of the water by photosynthesis, thus improve water quality. Details of some of the plants are as under.

Aquarium Plants

Aquarium plants provide many services to the aquarium. They create shade-hiding places, offering convenient safe retreats when danger threatens or when a fish simply wants to get away from it all. Plants also play a major part in keeping the aquarium clean.

Many fish make practical use of plants. The soft-leafed species make good eating for herbivorous fishes whilst the firmer, stout-leafed plants offer favourable spawning aids—either as spawning sites themselves or as nest-building materials.

Aquarium plants may be divided into three groups:

1. Floating Plants

2. Rooted Plants
3. Bunched Plants.

As their name suggests, floating plants are not anchored. In the gravel base but remain on the water surface; most have trailing roots which hang down in the water, absorbing nutrients and providing sanctuary for young fishes. The main group, rooted plants, makes up the decorative species and contains both fine and broad-leafed species. Bunched plants are rapid-growing species from which regular cuttings may be taken to be re-rooted in the gravel to provide extra stocks. The following list gives some suitable aquarium plants and their particular use.

(*a*) Floating Plants

e.g. Azolla caroliana, Ceratophyllum demersum, Echhornia azurea, Lemna minor, Salvinia aurucykata, Postia stratiotes.

(*b*) Rooted Plants

Propagation is with the help of roots. *e.g. Vellisneria spirilis, Echinodorus paniculatus, E. brevipdicellatus, E. brevipdocellatus, Aponogenton echinatus, Hydrilla verticillata, Ladwegia palustris, Sagittaria sublata.*

(*c*) Bunched Plants

e.g. Comba aquatica, Camboma caroliana, Egeria densa

Floating Plants

Azolla caroliniana, Fairly Moss, is a tiny fern, the leaves are red or green but often appear greyish due to the tiny hairs on their surface. Quite decorative but not of great practical importance as far as the fishes are concern.

Ceratophyllum submersum, Hornwort, is a cosmopolitan plant with hard bristly whorls of leaves. It is tropical cersion of the cold water species *C. demersum* and it prefers to grow as a tangled mass rather than be rooted down. The brittle leaves send out rootless very readily.

Lemna minor, Lesser Deckweed, can become a curse in the aquarium. Although its tiny oval leaves bring welcome shade, they spread with alarming haste to cover the entire water surface if unchecked. Some large Barbs relish them as food so netfils can be removed regularly to good effect.

Salvinia aurucykata, Butterfly Fern, is similar in many ways to Duck weed; as its hairy leaves are oval shaped, but larger, and it can be just as rampant. May be used for bubble nest building.

Postia stratiotes, Water Lettuce, resembles a floating lettuce except that the pale green leaves are velvelty, being covered in tiny hairs. When large, the leaves may easily become scorched by the aquarium lamps or damaged by condensation droplets, but the long trailing roots make very good retreats for young fry.

Rooted Plants

Aponogeton spp. do not have conventional roots but emerge from a rhizome or tuberous growth. Many are very beautiful, their leaves having ruffled edges or appearing lace-like where the tissue between the leaf veins is entirely absent. They bear flowers above water; pollinating these with a soft brush results in seeds being set which can be sown in shallow water to produce new plants. Requires winter rest period cooler water.

Cryptocoryne spp. make up a large family of very decorative aquarium plants. Leaf shapes can be narrow or broad depending on species. Sizes vary from tiny, gravel-carpeting species to huge "wedding bouquet" proportions. Most are quite happy to flourish in species lighting and the larger-leaved species make ideal spawning sites for Anglefish.

Sagittaria spp. are old aquarium favorites. Leaves are straps shaped and the plants are ideal for use as a background around the back and sides of the tank. Varieties of different sizes are available from low gravel converters to very large, water surface trailing types. It reproduces by young plants developing from vegetative runners although flowers may also be seen under aquarium conditions.

Vallisneria spp. are very similar in appearance to *Sagittaria* and can be used to the same effect. It is favourite and very decorative variety.

Vesicularia dubyana, Java Moss, is an excellent spawning medium for egg laying fishes. The tiny leaves provide a dense 'egg trap' as the spawning fish dive into it. Like cold water, Willow Moss it attaches itself to any firm surface by means of tiny rambling roots.

Branched Plants

Cabomba coroliniana, Corolian Fanwort, is a favourite amongst hobbyists, especially with those who find it no problem to grow. Its fine whorls of leaves need very clean water if they are not to become clogged with detritus. Bright light is also necessary for successful growth.

Myriophyllumo hippuriodes, Water Millfoil, is another fine leaved plant similar to *Cabomba* and requiring the same care, especially clean water and precention of clogging by algae.

Egeria densa, Giant 'Elodea', has three or four leaves which curl back from central stem at close intervals. A rapidly growing plant which provides many cutting which, in turn, send down root-lets to the gravel.

Details of all above plants are given hereunder:

Rooted Plants

Vallisneria spiralis

Type of Plant	:	Rooted Plant
Common Name	:	Tape Plant
Class	:	Angiospermae
Sub-class	:	Monocotyledonae

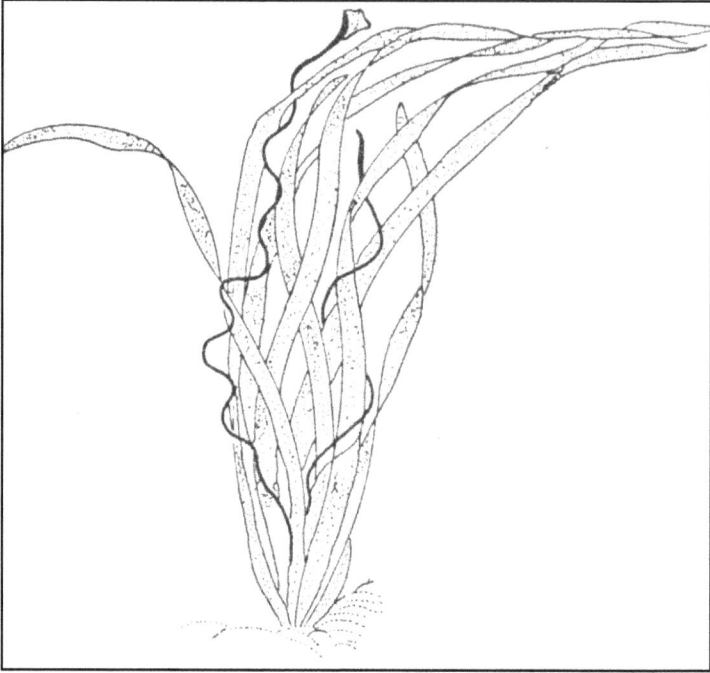

Figure 4.1: *Vellisneria spiralis*

Family	:	Hydrocharitaceae
Scientific Name	:	*Vellisneria spiralis* (Michx)

Identification

Vallisneria is dense plant with number of long, ribbon shaped and often spiraled several times around their axes. Leaves bear 3 to 5 parallel veins which run the entire length of the leaf. Steam is runner, found under ground bearing a rosette of leaves on the nodes at regular intervals. Males and females plants are separate. Male flowers are produced near roots and female flowers are seen with very long and coiled stalk. This is found in tropical and subtropical waters. Because of the structure of the leaf, they are commonly known as tape plant.

Biology

The plant spreads very soon in to dense clumps but is slow to establish in aquarium. The plant prefers neutral pH medium hardness and strong light. It requires soil with mixture of sand and clay for successful propagation. Plants are easy to grow. It can grow up to one meter. Propagation is easy by runner-cuttings.

Echinodorus paniculatus

Type of Plant	:	Rooted Plant
Common Name	:	Amazon sward plant

Identification

This is submerged aquatic plant found in tropical freshwater. Aquarium lovers know this species as Amazon sward plant. It is very beautiful aquarium plant with bright green colour. The leaves are tufted and lanceolate with veins running the whole length. Stem is a runner and the roots are very small.

Biology

Reproduction is by vegetative propagation through the off shoot which produces plantlets that grow independently after sometime. It requires clear water and good light. Growth rate is moderate compared to other aquatic plant. Substratum must be of coarse sand with rich nutrients.

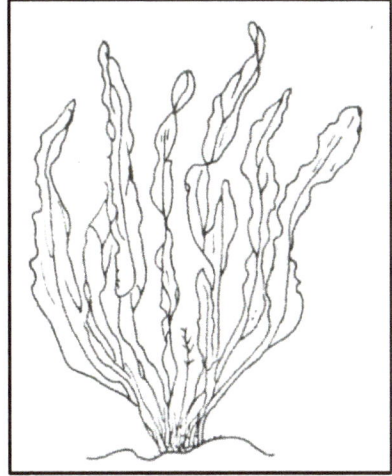

Figure 4.2: *Echinodorus paniculatus*

Cryptocoryne ciliata

Type of Plant	:	Rooted Plant
Family	:	Araceae
Scientific Name	:	*Cryptocoryne ciliata*

Identification

There are 20 varieties of *Cryptocoryne*. Plants bear attractive dark green lance-shaped leave with crinkled edges on long stalk. This makes them suitable for only deep-tank aquarium (50 cm in length) being creeper or runner (rhizone). They are propagated easily. Plants need frequent root cutting to keep a check on their fast growth. However, if planted in clean sand, they do not attain large size and remain small plants. They need 8–10 hours moderate illumination. The fast growing varieties are *C. affinis, C. griffithi,* and *C. willisti*.

Water Quality

They need slightly acidic water = pH 6.5 to 6.9. They can tolerate wide range of temperature *i.e.* 18° to 26° C.

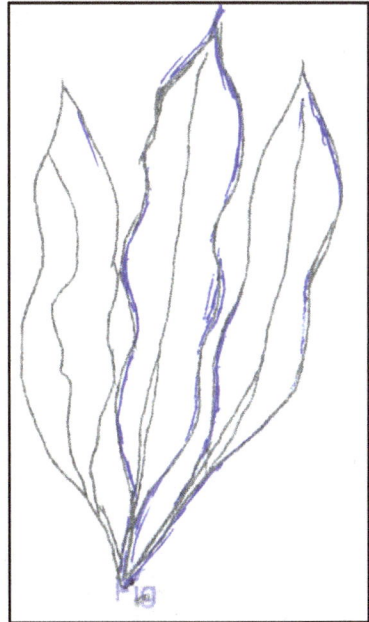

Figure 4.3: *Cryptocoryne ciliata*

Aponogeton echinatus

Type of Plant	:	Rooted Plant
Scientific Name	:	*Aponogeton echinatus*
Family	:	Aponogetonaceae

Identification

This beautiful plants are cultivated by rootstock, which is tuberous and cylindrical. The plant has very attractive lace leaves and flowers with crinkled edges. Cross-pollination occurs. Seed formation is difficult but plants propagate only by seeds. The plant survive for short period and die.

It is necessary to pull the bulb, take care of the root stocks so that these do not rot. After some time new leaves will appear. The bulb must be planted in rich compost. It is found in central India.

Other varieties are *A. crispus*, *A. ulvaceous* and *A. natuns* have bright coloured long wavy leaves in tuberous cylindrical rhizome. In this plant chlorophyll is absent and the entire leaf is formed to a pattern of leaf windows.

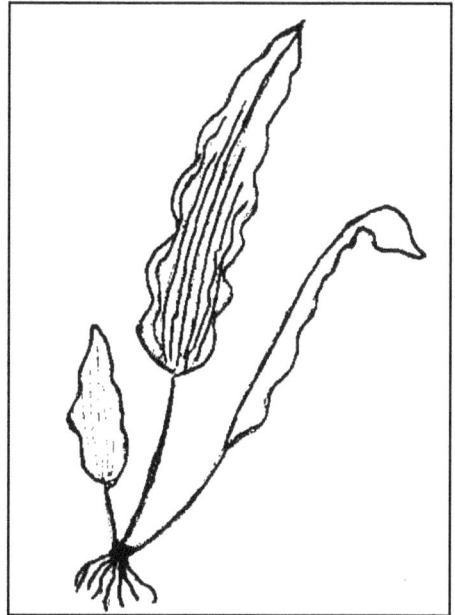

Figure 4.4: *Aponogeton echinatus*

Echinodorus brevipedicellatus

Type of Plant	:	Rooted Plant
Common Name	:	Amazone Sward Plant
Family	:	Alismataceae
Scientific Name	:	*Echinodorus brevipedicellatus*

Identification

These bright coloured plants are good piece of center of aquarium. It is propagated by runners on which develop small plantlets. Propagation is also from cutting of the runner. It needs 8–10 hours illumination and grows up to 50 cm in length. It reproduces fast and forms a carpet in aquarium. Generally it cultivated at the front side of aquarium. It is a flowering plant. The shortest stem is between 2 and 16 inches in length; topped by long, narrow lace-like leaves, oval shaped. In natural waters the plant will grow well above the surface, but in the aquarium they tend to lay flat along the surface. They grow faster at 21°C temperature.

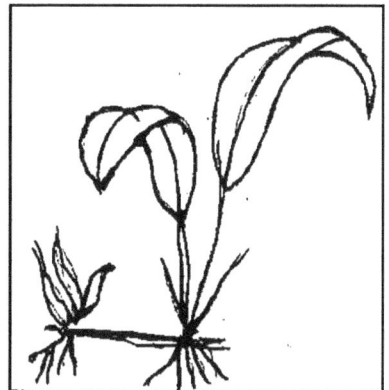

Figure 4.5: *Echinodorus brevipedicellatus*

Hydrilla verticillata

Type of Plant	:	Rooted Plant
Common Name	:	Hydrilla or Chain weeds
Class	:	Angiospermae
Sub-class	:	Monocotyledonae
Family	:	Hydrocharitacoae
Scientific Name	:	*Hydrilla verticillata*

Identification

Hydrilla is most suited for aquarium. The leaves grow to a size of about 2 cm and are found in whorls to 3–6 cm. The plant is profusely branched stem long cylindrical and branched leaves in group occur at short intervals forming whorls of several dark green small colour leaves. The roots are fibrous. This plant is found distributed in the tropical and subtropical waters. It grows submerged and mostly found with floating branches. In nature, it is a useful for the fishes to hide. It is a good oxygenator and easily propagated from cutting. It grows rapidly in a good light it is an excellent feed for some herbivorous fishes.

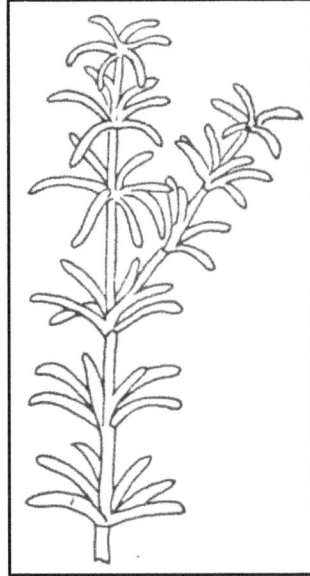

Figure 4.6: *Hydrilla verticillata*

Biology

Hydrilla can tolerate wide fluctuations in the growing medium. Reproduction is mainly through vegetative propagation. It is very easy to culture as it is fast growing species. They require strong light and pH between 7.8 and 8.0. It can be cultured in earthen ponds. This plant should be treated in 12 ppm $KMnO_4$ for 3–5 minutes before planting them in aquarium.

Ludwegia palustris

Type of Plant	:	Rooted Plant
Family	:	Anagraceae
Scientific Name	:	*Ludwegia palustris*

Identification

This is a flowering plant has leaves lying opposite each other, in pairs, alternately and at right angles, at small intervals. Its leaves are smooth oval in shape or elliptical, green to red above and red to violet below in colour. Petioles are weak. The trailing steam can produce roots at the nodes easily. The roots are strong and colourless.

Figure 4.7: *Ludwegia palustris*

Biology

Reproduction is by vegetative propagation. It is easy from top cutting. Growth is relatively fast. Requires soft water with nearly neutral pH. It shelters a number of fish species.

Culture

Ludwegia can easily be grown. The American of *Ludwegia* are in greatest demand of aquarium specimens. Intense illumination and rich soil are necessary to produce robust and healthy plants.

Bunched Plants

Caboma aquatica

Type of Plant	:	Bunched Plant
Scientific Name	:	*Caboma aquatica*
Family	:	Nymphaeaceae

Identification

Light green leaves are fan-shaped, growing opposite to each other at more or less regular intervals on rather long stem. Leaves are finely dissected in to 150–200 segments. Veins are not present on them. They are flowering plants, which are cultivated with difficulty. Young plants develop form shots arising at the base. This makes their propagation easy from cuttings. Plants prefer soft water.

Aquaculture Plants

Camboma caroliana

Type of Plant	:	Bunched Plant
Common Name	:	Fan Wort or Cabomba
Family	:	Nymphaeaceae
Scientific Name	:	*Camboma caroliana*

Identification

The slender stems bear coarse-cut leaves of these plants are arranged in fan shaped (half circled). These leaves are green in colour and brittle. Veins are not present on them. There is tendency for this plant to become stringy. Nipping the head will some times promotes busy growth. It is a good plant for garden pool. It needs bright light for 10–12 hours a day. It prefers soft water. pH 6.5 to 6.8 is ideal for cultivation of this plant.

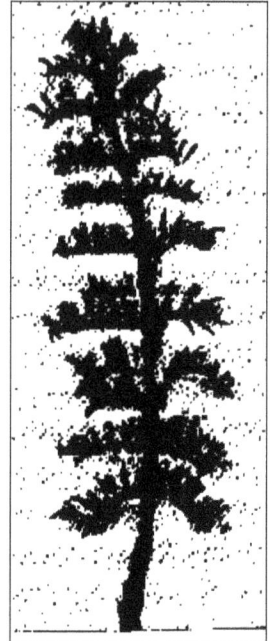

Figure 4.8: *Caboma aquatica*

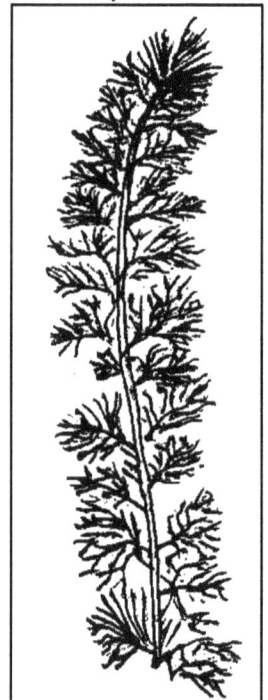

Figure 4.9: *Camboma caroliana*

Floating Plants

Ceratophyllum demersum

Type of Plant	:	Floating Plant
Scientific Name	:	*Ceratophyllum demersum*
Family	:	Ceratophyll-aceae

Identification

It is very fragile plant. Roots are absent. Plant has long branching stem bearing whorls of a few to 10 leaves. Each leaf is divided into 2–4 forked and spiky leaflets. Plants float freely submerged in water column and do not bear roots broken or cut pieces develop in to new plants. Growth is very fast. Frequent butting/thickening is essential to keep a check on their growth. Plants need good light and nutritious water. Plants prefer rather cool temperature. It provides good spawning medium to egg laying fishes and provides shelter to baby fishes.

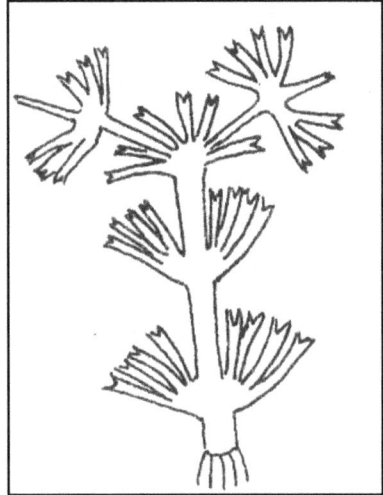

Figure 4.10: *Ceratophyllum demersum*

Echhornia azurea

Type of Plant	:	Floating Plant
Common Name	:	Blue water hyacinth
Family	:	Pontederiaceae
Scientific Name	:	*Echhornia azurea*

Identification

It is a flowering plant. It has submersed erect stalk. Stem is a modified rhizome leaves are light green but large and narrow and arranged alternately on stalk. The stalk when it reaches the water surface develops into emerged form with flat and round leaves. Only the submersed

Figure 4.11: *Echhornia azurea*

form is suitable for aquaculture. It is necessary to top the plant so as not to allow the emerged form to develop.

Azolla caroliniana

Type of Plant	:	Floating Plant
Scientific Name	:	*Azolla caroliniana*
Family	:	Azollaceae

Identification

The plant is a floating fern; leaves are pale green or reddish brown. They lie in two rows on a thallus; stem is branched. Roots are filamentous. They are fast growing, soon develop in to a blanket cover on the water surface. Cutting light to a great extent to adversely effect submerged plant.

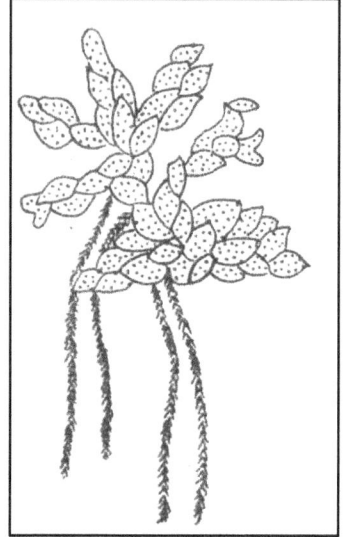

Figure 4.12: *Azolla caroliniana*

Chapter 5
Fish Feed and their Culture

Introduction

Fish needs better nutrition for obtaining better growth. In extensive aquaculture system the growing fishes utilizes the natural feed in the form of phytoplankton and zooplankton. But in semi intensive system natural feed is supplemented with artificial feed. On modern aquaculture emphasis has been given on intensive culture system in which secured fish production is envisaged on total artificially prepared well-balanced feed. Fishes are considered to be better-feed converters than the other vertebrates. However, in many parts of world at present semi intensive system of aquaculture practices are in vogue, due to higher cost factor involvement in intensive farming.

Fish Nutrition by Natural Feed

The fish originally lived solely on the natural feed of the water bodies. Live foods are mainly phytoplankton and zooplankton. For regular supply of live food, it is better to have species wise live food culture tanks.

The most common phytoplankton species used for culture are:

1. Chaetoceros
2. Chlorella
3. Tetraselmis
4. Dunaliela
5. Thalassiossira
6. Tubellaria
7. Rhizosolenia
8. Skeletonema.

Various enrichment Medias are used for the culture of desired species of phytoplankton by providing ambient conditions of light, Temperature, DO_2 Carbon Dioxide, pH, Salinity etc. The selections of culture containers are also species specific. The selection of ingredients for fertilizers depends upon the actual requirement of the individual species of algae. But, as far as culture of marine algae is concerned EDTA Na is considered to be an essential ingredient for better realization.

Micro Algae and Fish Food Organisms

Fishes feed on a wide variety of food items in the wild. In culture system, as far as possible natural foods are developed and the fishes stocked are allowed to feed. In addition, compounded supplemental feeds are also given. For rearing larvae in hatchery and for ornamental fishes and other fishes grown in confined conditions, the live food organisms are given by culturing *those organisms by scientific means.* The organisms, which are used to feed other cultivable organisms, are cultured in controlled and confined environments. Of the many organisms, the unicellular algae are included under Micro algae.

The unicellular plants in the aquatic system are termed as micro algae. Marine and freshwater micro algae are extremely important for the aquaculture industry. Majority of crustaceans' larvae feed on the micro algae. Larval and post larval stages of many fishes also require them as the stable hatchery feed. A number of marine micro algae are rich in poly unsaturated fatty acid (PUFA), which is vital in maturation of shrimps. Generally shrimp feeds are enriched by PUFA. Live food organisms like artemia and bivalves feed on micro algae and they are in turn fed to crustacean larvae. Besides, the micro algae serve as important source of feed for oysters and other bivalves.

Recently there is growing interest in micro algae for the bioactive compounds and industrial chemicals. For examples *Dunaliella salina* is rich source of carotene. Micro algae like *Spirulina* are being considered as food for human consumption as single cell proteins.

Artificial Feed

A diet for aquarium feeding should be nutritionally balanced, palatable and resistant to crumbling, water stable buoyant and inexpensive.

The diet of fish should normally have protein, carbohydrate, fat essential amino acids, vitamins and minerals in right proportions so that it meet the daily energy and nutrient requirement to support the maintenance and growth of fish.

Dry or flake feed is easy to manufacture, transport, store and use. Most nutrients in dry feed including vitamins and unsaturated fats are stable at room temperature and therefore such foods could be stored safely without freezing.

The dietary protein requirements are size dependent. *i.e.* small fish require higher level of protein for maximum growth than larger fish. Generally aquarium fish need feeds containing 30–40 per cent protein.

Ornamental fishes in captivity need to utilize dietary protein with utmost efficiency. The protein requirements of some ornamental fishes under captivity conditions are presented in Table 5.1.

Table 5.1: Protein Requirements of Ornamental Fishes

Common Name	Scientific Name	Dietary Protein Requirement %
Guppy	*Poecilia reticulate*	30–40
Gold fish	*Carassius auratus (Various strains)*	29–53
Tinfoil barb	*Barbodes alius*	41.7
Discuss	*Symphosidon aquifasciata*	44.9–50.1
Red head cichlid	*Cichlasoma synspilum*	40.81
Dwarf gourami	*Colisa lalia*	25–45
Swardtail	*Xiphophorus helleri*	45
Angel fish	*Pterophyllum scalare*	45

Source: Keshavanath P and Prakash Patil, 2006.

Different types of balanced diet prepared feeds are available in market. However, one can prepare themselves as per following methods. Carnivorous fish feed should have higher levels of fish meal and fish oil in order to improve its acceptability and increased intake.

An ideal 40 per cent protein feed suitable for aquarium fish should have the following ingredients

Formula I

	For herbivorous Fish per cent	For carnivorous fish per cent
Fish meal	25	30
Shrimp waste	22	10
Soya bean meal	25	20
Rice or wheat bran	20	25
Tapioca flour (as binder)	05	10
Fish oil	03	05

Formula II

1. Fish meal powder: 250 g
2. Soya bean balls: 200 g
3. Sago: 200g
4. Pearl rice: 250 g

5. Granular wheat flour: 200g
6. Egg (white): 10 Nos.
7. Vit B complex for + Biotin + Vit. C.: 20 tablets

Suitable Aquarium Foods

Introduction

Aquarium fishes are dependent up on the aquarist to provide them with a correct balanced diet, the overall effects of which will be seen in their colouration, growth, resistance to disease and their willingness to breed. Today, using the high-quality flaked and similar prepared diets that are available from pet and aquatic dealers most easily and conveniently provide a balanced diet for aquarium fish. Such foods have now largely replaced the old fashioned, inconvenient and often potentially dangerous to made and live foods were so popular. The live foods do still have a number of specific uses in fish keeping and these include;

1. Tempting delicate, fussy fish on to feed, before weaning them on to a prepared diet.
2. Conditioning adult fish for breeding perhaps by an abundance of live food acting as a "trigger" to initiate spawning.
3. Providing a useful first food for newly hatched fish fry, before they can take finely powdered dried food.

Live foods are important in the aquarium for providing trace element and boosting vitamins, although it may be that trill of the chase is the major reason for enhancing the fishes condition.

Live Food for Fry

1. Infusorians
2. Brine Shrimp
3. Sieved water fleas
4. Dafnia

Feed fish fry several times a day avoiding tank pollution, yet checking that the fry have full, rounded bellies. When on to powdered dry foods and other prepared diets as soon as practical, so as to maximize the growth rate of the fry.

Live Food for (Conditioning for breeding) Adult Fishes

1. White worm
2. Chopped earth worms
3. Water flees
4. Tubifex
5. Blood Worms

Food sparingly, perhaps several times a week use prepared foods as a staple diet, given sparingly two or three times a day.

Details of Live Food and its Culture Method

(1) Infusorians

Infusorians are tiny single-celled animals that abound in almost every body of water. They form an ideal first food for very tiny fish fry. They are easy to culture in large jars. To ensure a continuous supply, start a new culture every three or four days, until the fry will take a proprietary liquid fry food, brine shrimp or finely powdered dried foods.

Cultured Method

To a jar three-quarter filled with boiled, cooled tap water, add three or four brushed lettuce leaves, a whole banana skin, or even a little hay, which has had boiling water poured over it to break up the cells. Place the jar in a warm, moderately lit place with the lid off. Over the ensuring few days the culture should go cloudy and begin to smell slightly. Then it will clear as the infusorians develop. Once the culture is clear and "sweet smelling", pour or siphon it into the tank little at a time.

Obviously, it is important to time the availability of the infusorians to coincide with the fry, coming on to feed and then to maintain a satisfactory supply until the small fish can readily accept proprietary brands of prepared diets.

(2) White Worms

These worms of the *Enchytraeus* genus are approximately 0.5–1 cm long and are an excellent live food-whether chopped and fed to small fish or fed whole to larger fish. Starter cultures are available from most pet sores and white worms are quite easy to culture in a shallow box with tight-fitting lid.

Fill the box about three-quarters full with a good loamy garden soil. It may be necessary to add aquarium peat to a clay soil. The main objective is to choose an organically rich soil that will retain moisture. Just before adding the starter culture, spray the soil so that it is well moistened but not waterlogged. Add the starter culture and then a little moistened white bread, pre-cooked porridge or baby breakfast cereal as food. Place the lid on the box and store the culture in the dark at about 15–20°C. Provide ventilation for the culture by way of small air holes in the sides of the box or in the lid.

Inspect the culture from time-to-time, paying particular attention to the moisture content. If the soil dries out, the worms will be unable to survive. But do not overfeed the culture; the uneaten food may 'sour' the culture and allow pests to flourish. Experience will indicate the correct amount of food to be provided: as a guide each new batch of food should be eaten within two or three days. Remove any uneaten food as you add fresh supplies.

Spread out the food in small pieces; the white worms will collect directly beneath these islands and be easy to remove with forceps. If it proves difficult to separate the worms from the food or soil, simply place a spoonful of the 'mix' in a saucer and cover it with water. The worms should migrate outwards and be easy to collect for immediate use.

Within about a month of setting up the culture in acceptable conditions, the population of white worms should approximately double. Within two months, a single culture should be providing enough worms for a reasonable collection of fishes.

Culture of White Worms (*Enchytraeids*)

These are soil dwelling creature and very similar to earth worms. White worms cultures are available in winter as supplementary food, they are easy to bread, they provide food that is highly nourishing. The worms obtained from stores (Market) are placed in the depression and mesh spread over them; they multiply at temperature near 15°C and when the soil is wet. After boxes are set up, rotate them for feeding the aquarium fish. They may get culture time to reproduce and ensures an adequate supply. Before feeding, worms should be free of soil particles and give quick rinse in water. This will remove any adhering particles or take a few and place them in shallow container such as food container lid and place where it can receive a gentle heat. Worms will amalgamate into a ball for making collection easy (Amita Saxena, 2003).

(3) Earthworms

An excellent and often overlooked live food is the common earthworm. Although you buy these from some pet stores and angling tackle dealers, if you have access to a garden, allotment or patch of waste ground then you should be able to collect more than a sufficient supply for your fishes. In damp weather, simply dig them up, or pick them up when they come to the surface of the lawn after an evening shower of rain in the summer. In dry weather, place one or two damp sacks in a shady part of the garden and bait them with some potato peeling or similar vegetable scraps about once a week. You will also find an abundant supply of earthworms around manure heaps in farms or stables.

After collection, keep the earthworms for a few days in sealed container-with small air holes for ventilation-containing a little damp grass or moss. During this time, they will clean themselves of soil etc. and will then be more palatable to the fish. You can use earthworms whole or chopped depending on their size and the size of fishes you are feeding.

(4) Sludge Worms

Tubifex and other similar tubificid worms are a familiar form of live food. These slim, maroon worms, about 1 cm long, are useful for tempting fish such as discuss on to feed, and as a live food for adult breeding fish. *Tubifex* is not easy to culture, however, and so it is best to buy supplies as you need them from your local aquatic store.

Use *Tubiflex* sparingly in the aquarium, and as an occasional food rather than a staple diet. Before use, gently rinse the worms in cold, running tap water for several hours, and perhaps gives a preventative treatment with one of the liquid food disinfectants available from aquatic stores. Once cleaned, you can keep *Tubifex* worms alive for sometime in a shallow dish of cold water, flushing it through with fresh water every day.

(5) *Euglena*

Culture of *Euglena*

Euglena is found in abundance in pond water, having rich nitrogenous waste matters. For culture take a medium sized battery jar and fill it with water and add to it 100 gms of wheat or rice bran and some hay, Keep this solution near the window so that sunlight do not fall directly and leave it for seven days. Now add some pond water containing *Euglena* and watch it under microscope. In two weeks, the water becomes greenish brown and surface of water will be covered with scum containing large number of *Euglena*.

(6) *Isochrysis galbana*

Classification

Division	:	Chlorophyta
Class	:	Chrysophyceae
Order	:	Chrysimonodales
Family	:	Isochrysidae
Scientific Name	:	*Isochrysis galbana*

Salient Feature

This is a unicellular, motile alga with two flagella, which are equal in length. Body is covered with circular scales arranged in several layers. Benthic stages consist of aggregation of cells, surrounded by a thick m. It is marine form, which is planktonic. Size ranges between 5 to 20 microns.

Figure 5.1:
Isochrysis galbana

These are among the smallest category of phytoplankton. They are also known as *Coccolithrophores*. They are having soft bodies which are shielded by tiny calcified circular plates. These are normally found in the open sea, but their profuse occurrence has been recorded in coastal water. They form important diet components of filter feeding animals.

Biology

Reproduction is by asexual method. They multiply in large number by longitudinal division in motile stage within two days.

Culture

In order to culture, calcareous elements may become deposited in the matrix and these resemble a nanofosil, known as "Tetralithus", which is simpler than cocoliths. This species is being mass cultured in controlled condition in nutrient media for use as larval feed in finfish and shellfish hatcheries. To one liter of seawater, 10ppm of KNO_3, 10 ppm of K_2HPO_4, 5 ppm of NA_2SIO_3 and 10 ppm of Sodium salt of EDTA are added and pure culture of *isochrysis* is inoculated and culture can be seen after 4 days.

(7) *Spirulina spirula* (Turpin)

Classification

Division	:	Cyanophyta
Class	:	Cyanophyceae
Order	:	Oscillatoriales
Family	:	Oscillatoriacea
Scientific Name	:	*Spirulina spirula*

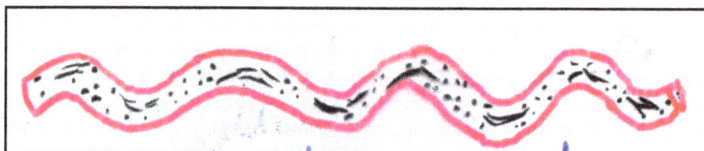

Figure 5.2: *Spirulina spirula*

Salient Feature

This is one of the simplest filamentous algae, which is helical and differentiated by the terminal cells, which differ from other cells of trichome. Transverse walls obscure under light microscope. *Spirulina* is promising food source for aquatic animals and terrestrial animals even to human beings. Every health food constitute *Spirulina* and several tons are produced daily as it is a good source of protein. It's protein content ranges from 45–49 per cent in dry weight *Spirulina* inhabitats freshwater bodies in tropical and temperate regions.

Biology

Reproduction by hornogonium formation.

Culture

This micro algae is one of the species cultured in large scale in ponds for use in the aquaculture industry and as a source of protein in food for other animals. Now-a-days it is being used as human food and medicine and hence mass cultured.

Importance of Spirulina

This simple cyno bacterium has been promoted as a health and sliming food. The United Nations' World Food Conference declared Spirulina as the best food for tomorrow as 1 kg of Spirulina consists as much nutrition as of 1000 kg of vegetables.

This blue-green alga contains proteins, amino acids, vitamins, chlorophyll xanthophylls, phycocyanin, beta-carotene, iron, magnesium, potassium, poly unsaturated fatty acids and little cholesterol.

Uses

Spirulina increases lactation in nourishing mothers; existing natural formulations get a boost with Spirulina. Ayurvedic lehlas aristas and churns are made to have Spirulina. It's having no side effects and non-habit forming characteristics find it a place in modern medicines too. This lowers blood sugar levels, cures pancreatitis, hepatitis, Cirhosis, glaucoma, cataracts, gastric ulcers, night blindness, liver disorders, anemia etc. Most of all, it is general health medicine. Fish fed with Spirulina put-up weight. Ornamental fishes get bright colouration. It is fed to silkworm to increase silk yield.

(8) *Dunaliella salina* (Teodoresco)

Classification

Division	:	Chlorophyta
Class	:	Chlorophyceae
Order	:	Volvocates
Family	:	Polyblepharidacea
Scientific Name	:	*Dunaliella salina* (Teodoresco)

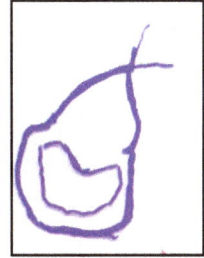

Figure 5.3:
Dunaliella salina

Salient Feature

The cells are naked and contain single chloroplast with eyespot and have single nucleus. Contractile vacuole is absent. There are two equal apical flagella, which characterize this species. It can form smooth walled gelatinous cysts. Glycerol is its major photosynthetic product. The glycerol concentration range from 0.5 M to 5 M. This species has industrial potentials in the production of glycerin and p-carotene. This species occurs in extremely saline habitats such as salt-pans and sea side rock pools. It can survive in more than 200 ppt salt concentration.

Biology

Reproduction is by both sexual and asexual methods. In asexual reproduction, the cells divide longitudinally. In the motile cells, the sexual reproduction is isogamous and meiosis occurs at zygote germination and results in the formation of motile individuals.

Culture

This species is being mass cultured for the production of glycerol and p-carotene in high saline areas. Increasing the selectivity of the cellular medium can increase carotene content in the cells. For Artemia species, which is a very important live food organisms for the successful operation of shrimp hatchery, this is the main food organism in the salt-pans.

(9) *Chaetoceros affinis* (Lauder)

Classification

Common Name	:	Chaetoceros
Family	:	Bacillariophyceae
Scientific Name	:	*Chaetoceros affinis* (Lauder)

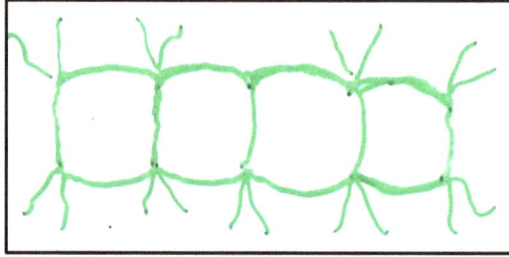

Figure 5.4: *Chaetoceros affinis*

Principle

In shrimp hatchery, one of the most suitable diatom is Chaetoceros spp. It is preferred because diatom have cell dimension ranging between 10–24 micron shrimp larvae in protozoa stage can consume this with their tiny mouth part. Secondly, the nutritive value of *Chaetoceros* spp. is also fairly good.

Availability

In nature, the availability of Chaetoceros cells is uncertain due to seasonal variation. However, they are available in Gulf of Kutch during December to March but in summer and monsoon it occurs occasionally.

Aim: Mass Culture of Centric Diatom *Chaetoceros affinis*

Principle

In shrimp hatchery one of the most suitable diatoms is *Chaetoceros* spp. It is preferred because diatom have cell dimensions ranging between 10–24 micron shrimp larvae in protozoa stage can consume this with their tiny mouth.

Material

1. Sodium nitrate: 16 mg/liter
2. Sodium silicate: 8 mg/liter
3. Potassium dihydrogen orthophosphate: 8 mg/liter
4. EDTA NO_2: 8 mg/liter

Method

(A) Stock Solution

Dissolve above materials in one liter of distilled water.

(B) Culture

The sample of seawater should be collected during the income tide. Filter seawater using 100 micron mesh filter cloth. In receding tide number of Chaetoceros cells are less.

Tank with 400 liter water capacity to be filled with filtered sea water and aeration should be given. Keep the tank in open sun light. The isolate cells of Chaetoceros may be added in the aerated tank. The enrichment media, which was prepared as a stock solution, is to be added at the rate of 1 ml/liter.

Aeration should be continued. When water colour is changed to greenish all seawater along with Chaetoceros cells should be transferred protozoa to shrimp larvae for feeding.

(10) *Chlorella salina* (Butch)

Classification

Division	:	Chlorophyta
Class	:	Chlorophyceae
Order	:	Chlorococcales
Family	:	Oocystacea
Scientific Name	:	*Chlorella salina*

Figure 5.5: *Chlorella salina*

Salient Feature

This is a unicellular alga. The cell is spherical in shape. Cell wall is smooth and moderately thick throughout. There is only one parietal chloroplast. *Chlorilla salina* is used in shrimp, Pearl Oyster and Edible Oyster hatcheries. It is also used as food of *Brachionus* sp.

Biology

This unicellular alga is marine and planktonic. The reproduction is by asexual method.

Aim: Mass Culture of *Chlorella salina*

Principle

Chlorella are phytoplankton and they are used as food in the culture of Rotifer (*B. plicatilis*).

Method 1

Material

1. Ammonium sulphate: 100 g/1000 liter
2. Single super phosphate: 10 g/1000 liter
3. Urea: 10 g/1000 liter

Procedure

Fertilize 1000-liter water using above media. Inoculate with pure culture of *Chlorella*. As *Chlorella* reaches its peak density, (10×10^6 to 20×10^6 cells/ml) take out 50 per cent of *Chlorella* for giving to rotifers. Tank is re-fertilized with the same media as was used initially. In order to main long-term culture in the same tank, about 25 per cent water is replaced from bottom with fresh filtered water at an interval of every 5 days.

Method 2

Chemicals

KNO_3, Na_2H_2O, $CaCl_2$, $6H_2O$, $FeCl_2$

Preparation of Culture Media

Solution–A: Dissolve 20.2 gm KNO_3 in 100 ml distilled water.

Solution–B: Take 4 gm of Na_2H_2O, 2 gm of $CaCl_2$ and 2 gm of $FeCl_2$. Dissolve in 80 ml distilled water.

Take 6 ml of solution–A and add 3 ml of solution–B. In this medium add pure culture of *Chlorella salina* and expose to Sun light.

After 5 days the *Chlorella* population density reaches 30 million cells/ml.

(11) Infusoria

Classification

Phylum	:	Protozoa
Class	:	Ciliata
Scientific Name	:	Infusoria

Introduction

Infusoria is the highest sub-division of protozoa, but we may use the term for a minute animal life on which young fish feed, also referred to so plankters. The first feeding should be of *infusoria*. Plenty of them are found in all mature water. But an aquarium has to be supplemented with *infusoria* and hence their culture to be maintained separately.

Controversy should not be raised whether which method of culture is best. Generally *infusoria* grow on vegetable matter. Bruised lettuce leaves, sliced raw potato, spinach, chopped dry hay, uncooked lentils etc, have all been lentils found successful for the purpose. The matter should be reduced to pulp, stirred into water drawn from a pond, which has no *Daphnia* and allowed to stand for couple of days. With luck plenty of *infusoria* may develop, through not guaranteed, expect foul smell. Culture of *infusoria* must be kept fresh and aeration helps a lot. Some decomposing vegetable matter may do feeding of *infusoria*.

Post larvae of fish eat very large quantities of *infusoria* and this is the secret of raising young fish to maturity is to keep them feeding from dawn to dusk. The drip method of feeding is the simplest and thus best. A jar containing healthy culture of plankter is raised above the aquarium. A length of ¼ inch of rubber tubing is put into the jar and aquarium as siphon. And the run of culture from jar is regulated to a steady drip by means of pinch cork. Fresh cultures are supplied in the same way as and when required. Other method is to add feed in, aquarium few teaspoon, full of culture every 2–3 hours. Irrespective of way feeding two precautions are necessary, firstly, the culture must be at the same temperature as that of aquarium and secondly, the culture must be strained/filtered through very fine muslin. If these precautions are not taken, high mortality may result.

Live bearers are about a quarter of an inch when born and so take larger food than *infusoria*. Most other, however, depends on *infusoria* for 3–4 weeks, are should be taken for young ones of species with small of mouth (*e.g. Colisa lalia*–Dwarf Gaurami) they may die while consuming larger *infusoria* by chocking. (Amita Saxena, 2003)

Successful production of seed of finfish and shellfish species in aqua hatcheries mainly depends on the supply of abundant quantity of proper live food organisms at appropriate time.

Infusoria are most primitive of all organisms in the animal kingdom. Besides being small in size, they are soft bodies and nutritionally rich.

Habitat

Infusorian micro-organisms is habitats ponds and tanks of freshwater, brackish water and marine habitats having decaying weeds, organic matter and foul smelling debris.

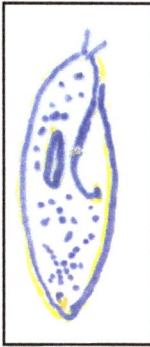

Figure 5.6: Paramecium   **Figure 5.7: Stylonychia**

Culture of *Paramecium*

Paramecium is found in fresh water rich in decaying organic matter. The culture of *Paramecium* is one of the easiest among protozoan as it reproduces very rapidly. The culture of *Paramecium* take 15–20 gains of wheat and some hay and boil them in 500 ml of water for a few minutes. All are allowed to cool down. Afterwards collect some, and water having submerged leaves containing *Paramecium*. Add the pond water to culture solution. In few days *Paramecium* will appear. The best temperature to culture in is 80–85°F.

Identification

They are tiny microscopic single celled animalcules.

Food and Feeding

Infusoria feed upon the micro organisms such as bacteria, algae flagellates and also on debris. Cilia present on the body act as chief locomotors and food catching organelles in most of the infusoria.

Reproduction

Two types of reproduction occur in infusoria *i.e.* asexual sexual. Asexual reproduction occurs by binary fission and sexual reproduction by conjugation.

Aim: Culture of Freshwater Infusoria

Material

Two glass aquaria of 50 liter capacity

Mosquito net cloth

Three banana peelings

10 ml of milk.

Method

Thoroughly clean an aquarium and keep in a cool place where natural light is available. Fill with 40-liter freshwater. Keep three bananas peeling in the aquarium. Add 10 ml of milk. Cover the aquarium with mosquito net cloth to avoid flies and mosquitoes. Leave the aquarium undisturbed; within 2 days water turns milky and emit foul smell. A film of slime forms on the water surface. In about 4–5 days water turns clear and become transparent with light yellow colour which is because of floating spores of infusoria have settled on water are feeding on bacteria and multiply in large numbers. Gradually film of slime on water surface break up and disintegrate. The culture is ready to feed the larvae of finfish and shellfish in aqua hatcheries.

Calculations

Thoroughly agitate the aquarium and take one ml of sample and preserve in 5 per cent formalin solution. Estimate infusoria with help of "Sedguick–Rafter" counting cell by using following formula:

$$N = \frac{a \times 1000 \times 2}{L}$$

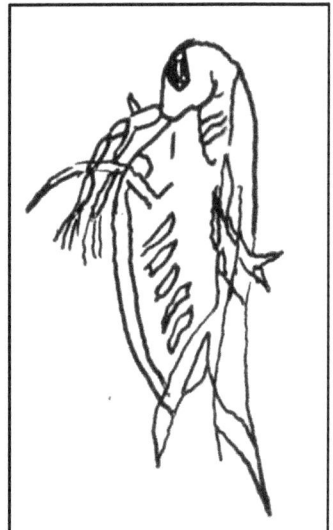

N: Number of Infusoria

a: Average number of infusoria in all counts in counting cell of 1 mm^3 capacity.

c: Volume of original concentration in volume.

L: Volume of original water.

(12) Cladocerans

Classification

Common Name	:	Moina, Water fleas
Phylum	:	Arthropoda
Class	:	Crustacean
Sub Class	:	Branchlopoda

Figure 5.8: Moina

Order	:	Cladocera
Family	:	Daphnidae
Scientific Name	:	*Moina micrura*

Introduction

They are commonly known as '"water flaws". They are commonest reproductive of this group of crustaceans. They are very popular. This is easily available in nature and easy adaptable to captive conditions.

Habitat

Moina inhabit fresh and low saline water ponds, tanks, and some sea wage lagoons.

Identification

Moina measures 0.5 to 1.0 mm in length and 0.2 to 0.6 mm in width. Moina has a pair of prominent caudal setae. Head is round and bears a pair of large biramous antennae, a pair of small antennules and a compound senile eye. Large biramous antennae are the chief organs of locomotion. Thorax bears 5 pairs appendages.

Food and Feeding

They feed on algae, fungi, protozoans and organic debris.

Reproduction

Generally Moina reproduce parthenogenetically. The eggs are laid in large brood pouch situated between abdomen and posterior carapace. The eggs undergoes complete development in brood pouch chamber before being released as first instar which is similar in morphology as that of adult female. A batch of eggs in brood chamber is termed as "brood". The young ones are released in small batches known as "clutches". In the total life span of a parthenogenetic female, a sexual phase occurs by generating sexual males. These males after mating with parthenogenetic females turn to sexual female, which results in production of resting eggs known as "ephippala". This ephippala can store for initiating the fresh culture as and when desired. They remain in viable condition for about 2 to 3 months.

Aim: Culture of *Moina micrura* with Locally Available Organic and Inorganic Fertilizer

Material

250 micron bolting silk cloth.

Plastic bucket

Microscope

Dropper, Slides, Test tubes of 50 ml capacity.

250 ml beaker and 1–2 lit Jar.

500 lit capacity cement tank.

Chemicals/Ingredients
1. Ground net oil cake
2. Single super phosphate
3. Urea
4. Ammonium sulphate
5. Chicken manure

Preparation of Stock/Pure Culture

Method (Stock Culture)

Collect *Moina micrura* using 250 micron filet cloth along the shore of pond collect sample early morning hours. Observe under microscope. Pick Moina with the help of a dropper. Inoculate 2 or 3 moina in each 50 ml test tube filled with 20–25 ml of water. Feed Moina with chlorella or 200 ppm yeast. After 24 hrs observe the test tube and count off springs. Each Moina gives 6–10 of spring in 24 hrs. Put this culture in 250 ml beaker and feed as above. After 48 hrs these culture may be used for mass culture.

Mass Culture

Thoroughly clean the 500 lit capacity cement tank. Fill the tank with 400 liter of filtered freshwater. Arrange mild aeration. Fertilize tank with 40 g ammonium sulphate, 4 g of single super phosphate and 4 g Urea. After fertilization inoculate tank with chlorella or mixed algae. After 72 hrs observe algal cell density. If cell density is $20^{-3} \times 10^6$ cells/ml. inoculate Moina stock culture @ 40^{-5} individuals per liter. If algal concentration is reduced, re-fertilize the tank with 75 ppm of groundnut oilcake at an interval of every 4 to 5 days. Moina attains a peak density of 20,000 to 25,000 in about 5 to 7 days after inoculation stop aeration, wait for 10 minutes and harvest Moina early in the morning with the help of scup net made of 250 micron cloth. After washing it can be given to the larvae of fish.

(13) *Brachionus plicatilis*

Classification

Common Name	:	Rotifers
Phylum	:	Rotifera
Class	:	Monogononta
Order	:	Ploima
Family	:	Brachionidae
Sub-family	:	Erachioninae
Scientific Name	:	*Brachionus plicatilis.*

Introduction

Rotifers are commonly called as "Wheel animals". They are usually microscopic. Among the rotifers, *Branchionus* spp. become more popular as "Live Food", because

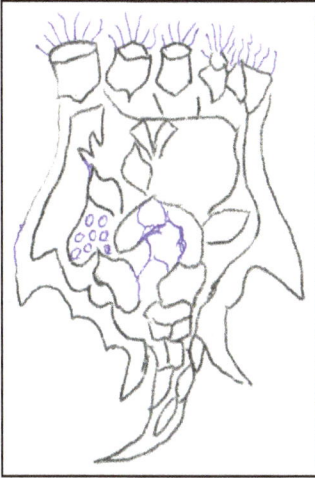

Figure 5.9: Female Rotifer

Figure 5.10: Male Rotifer

of its high nutritive value, small size, worldwide distribution, fast multiplication and easy adaptability to captive culture. *B. plicatilis* is widely used as Prime live food for early stages of fish and invertebrates in many aqua hatcheries. It is slightly larger than infusoria. It can also be used as a first food for most of the zooplankton feeding larval stages of fish.

Habitat

The pelagic herbivore rotifer *B. plicatilis* in a wide range of water bodies such as brackish water, salt, lakes and backwaters. It occurs in tropical and sub tropical waters all over the world *B. plicatilis* thrive best when temp range is between 22–25° C and salinity between 10 to 15 ppt.

Size

The size of amitotic *B. plicatilis* varies between 150–250 micron in length (without foot) and 100 to 150 micron in width.

Food and Feeding

B. plicatilis is fastidious filter feeder, feed on particle size. It prefers food items which are simple in shape and well suspended in water. It obtains food through ciliated corona. The most tested and acceptable food for *B. plicatilis* are *Chlorella* spp. and *Tetraselmis* spp. with the addition of bakers/marine yeast. Maintenance of high food concentration helps in producing maximum yield in the culture system.

Reproduction

B. plicatilis undergo two types of reproduction depending upon the culture condition. Under favourable conditions, the most predominant mode of reproduction is parthenogenesis (amitotic reproduction). However, in unfavourable condition, it restores to sexual reproduction (mystic reproduction). Under favourable conditions each parthenetic female produce about 16 to 20 amitotic eggs in its life spawn of about 5 to 7 days. The doubling time varies between 6 to 18 hrs depending on the availability of quantity and quality of food in the culture system.

Culture: Stock culture

In order to start stock culture collect *B. plicatilis* from the stagnant salt water bodies (salinity between 10 to 40 ppt) with the help of a scoop net having 50–100 micron mesh. Dilute the sample by adding freshwater having same salinity. Feed the V with yeast @ 200 ppm or Chlorella at a cell density of 10×10^6 cells per ml.

Aim: Culture of Rotifers

Principle

Rotifers are microscopic Zooplankton. They are popular as "Live food" because of its high nutritive value, small size, worldwide distribution, fast multiplication and easy adaptability to captive culture.

Material

1. Test tubes
2. 100 ml beakers
3. Chlorella or
4. Yeast

Method

For pure culture, collect *B. plicatilis* from the stagnant salt-water bodies (Salinity range between 10 to 40) with the help of scoop net of 50–100 micron. Examine under microscope and with the help of fine dropper pick up *B. plicatilis* and culture in glass tube containing 5 ml of water. Feed the *B. plicatilis* with yeast @ 200 ppm or Chlorella at the cell density of $10–10^6$ cells per ml. Serially dilute the test tube culture daily to test tubes of 20 ml capacity containing 10 ml water. Gradually increase volume to 50 to 100 ml capacity beaker.

(14) *Tubifex*

Classification

Common Name	:	Tubifex worm or Sludge worm
Phylum	:	Annelida
Class	:	Chaaetopoda
Order	:	Oligochaeta
Family	:	Tubificidae
Scientific Name	:	*Tubifex tubifex*

Introduction

It is good live food source and can be obtained all over the year. Their length varies from 3/8 inches to 2 inches long. They are extremely thin and rusty red in colour and found in plenty of mud, streams and pounds, collected by lifting the mud with a spade and putting it into muslin bags. Then swell the bags in streams remove all the worms. They form a ball like mass in shallow container. Worms are stored in a bucket or large bowl in to which a slow steady streams of water is allowed to enter. If any time worms become bunched jet of water is poured for separating them. A bunch is dropped in to aquarium, they will eventually scattered and

Figure 5.11: Tubifex

bury some [part of themselves in the gravel, the protruding ends waving round as if in strong breeze. Fish eating these individual swarms usually only manage to bite in the gravel to decay and cause problem later. It is, therefore, advised to keep these worms in glass bowl and they may not scattered.

Some of the micro-worm forms staple food for the post-larvae, juveniles, adults and brood stock of finfish and shellfish. Among them *Tubifex* larvae are commonly used as live food for the maintenance of ornamental fishes.

Identification

Tubifex worms are long and slender. There is no distinct head. At the extreme anterior end there is a lip like structure known as the "Postomium", which is not counted as segment. The mouth is present in the first segment. Body of the animal is thread like and cylindrical in cross section. These worms are generally found in place where heavy settlement of organic materials and degradation and decomposition takes place and in streams where organic pollution occurs. Part of its body is found rising vertically and wriggling in a wave like fashion of motion. Even in the slightest disturbance, they quickly withdraw into the soft sediment. After sometimes, they come again to the surface of the sediment and wave the body and breathe available dissolved oxygen. When the worms are segregated from the sediments they form a ball.

Habitat

They are also abundantly found in places where suspended material from sewage settles in streams of reduced flow rate. These areas are known as "Sludge banks". *Tubifex* is one of the most reliable indicators of organic pollutions. These worms live partly buried into the sediments.

Food and Feeding

Tubifex worms feed on decaying organic matter, detritus, vegetable matter which commonly available in sewage drains.

Reproduction

Tubifex worm is hermaphrodite, because it has both male and female organ in the same animals. In nature specimens, the reproductive organs are clearly found on the ventral side of the body.

Both organs become mature at different times, and thus self fertilization is avoided. The fertilized eggs undergo complete development in the cocoon by utilizing the aluminous nutritive fluid for growth.

Culture

Tubifex can be easily cultured on mass scale in containers with 50 to 70 mm thick pond mud at the bottom, blended with decaying vegetable matter and masses of bran and bread. Continuous mild water flow is to be maintained in the container, with a suitable drainage system. Within 15 days cluster of *Tubifex* worm develop and this can be removed with mud mass.

These worms are considered as one of the best, cheap and most easily available live foods which can be fed to fin fishes, prawns, shrimps and frogs. They are cultured in commercial scale and sold to aquarist to feed graceful ornamental fishes and other fish culturist. The worms grow fast on a substance containing 75 per cent cow dung and 25 per cent fine sand with continuous running water at the rate of 250 ml/min. in a culturing system of $150 \times 15 \times 15$ cm. Addition of cow dung at the rate of 250 mg/ cm^2 once in 4 days give optimum growth and the worms can be harvested at a rate of 125 mg/cm^2/30 days. The worms can be harvested before dawn or after dusk since they are photophobic.

Storage

A bunch is dropped in to aquarium, they will eventually scattered and bury some [part of themselves in the gravel, the protruding ends waving round as if in strong breeze. Fish eating these individual swarms usually only manage to bite in the gravel to decay and cause problem later. It is, therefore, advised to keep these worms in glass bowl and they may not scattered.

(15) Daphnia

Classification

Common Name	:	Water Fleas
Phylum	:	Arthropoda
Class	:	Crustacea
Sub Class	:	Brachiopoda
Order	:	Cladocera
Family	:	Dephnidae
Scientific Name	:	*Daphnia* spp.

Introduction

Daphnia are the commonest representative of group crustaceans. They are popular as live food in aqua hatcheries. This is because of their easy availability in nature and easy adaptability to captive conditions.

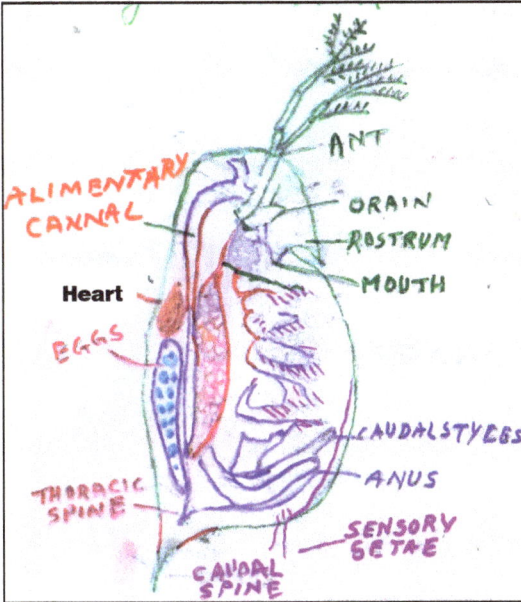

Figure 5.12: Daphnia

Morphology

Body is laterally compressed and enclosed in a bivalved shell or a large fold of carapace. *Daphnia* is distinguished from Moina by the presence of prominent caudal spine. *Daphnia* has small setae. Head is round and bears a pair of large small antennules and a compound sessile eye. Large biramous antennae are the chief organs of locomotion. Thorax bears 5 pairs of appendages.

Size

Daphnia measures 0.5 to 2.5 mm in length and 0.3 to 1.0 mm in width.

Food and Feeding

Daphnia feed on algae, bacteria, fungi, protozoan and organic debris.

Reproduction

Daphnia reproduce parthenogenetically. The eggs are laid in large brood pouch situated between abdomen and posterior part of carapace. The eggs undergo complete development in brood chamber before being released as a first instars which is similar in morphology as that of adult female. Eggs in brood chamber is termed as "Brood". The young ones are released in small batches. In the total life span of a parthenogenetic female a sexual phase occurs by generating sexual males. These males after mating with parthenogenetic female turns to sexual female which results in production of resting eggs known as "ëphippia". These ephippia can be stored for initiating the fresh culture as and when desired. They remain in viable condition for about 2 to 3 months.

Culture of *Daphnia*

Stock Culture

Disease free stock of *Daphnia* should be collected from fish free pond using scoop net having 250–500 micron mesh size. Keep in plastic bucket. The sample is then diluted by adding fresh clear water and examine under microscope. Pick up *Daphnia* with the help of fine dropper. Keep 5–6 *Daphnia* in 20 ml glass tube containing 10 ml of filter water. Put yeast @ 200 ppm in test tube. Each *Daphnia* will produce 8 to 10 off spring in about 24 hours. Slowly increase volume to 100 ml beaker and feeding of yeast may be given in the same ratio. In the beaker culture of pure *Daphnia* can be observed.

Mass Culture

For culturing *Daphnia* use an old aquarium or tub containing about 20 to 30 liter of tap water that has been allowed to stand for 48 hours. Before adding the "Standard Culture" of water fleas- obtained from aquarium store-fertilize the water with ground nut oil cake (75 ppm), Single super phosphate (20ppm), and Urea (8ppm), with Chlorella or mixed phytoplankton. When algal bloom are develop within 3–4 days, *Daphnia* is inoculated @ 40–50 individuals per liter of water. It attains peak density within 5 to 7 days after inoculation. Now you can harvest them using a fine meshed net several times a week to feed to aquarium fish fry or to condition adult fish for spawning at one time. *Do not remove more than about 20 per cent of the Daphnia for feeding purpose.* Since this may deplete the culture beyond recovery.

You can culture *Daphnia* out door in plastic or glass covered containers during the warmer months of the year, although they will go dormant during the winter. Alternatively culture *Daphnia* indoor at temperature above 20°C and with some overhead illumination. This will provide a supply of live food through most of the year.

Enrichment

Daphnia form very cheap source of live food for many culture able species of finfish and shellfish. Enrichment of these organisms with omega-3 essential fatty acids helps in improvising its nutritional status. Micro algae, which are rich in omega-3 fatty acids, are fed to the *Daphnia* to increase food value. Nutritional quality of *Daphnia* can be accomplished with respect to Vitamin-C supplementation. Vitamin-C is included in the culture medium at the rate of 0.5 to 1.0 ppm. Vitamin-C builds resistance to diseases and stress during larval rearing in aqua hatcheries.

(16) *Artemia salina*

Classification

Common Name	:	Artemia
Phylum	:	Arthropoda
Class	:	Crustacea
Sub Class	:	Brachiopoda

Order	:	Anostraca
Family	:	Artemidae
Scientific Name	:	*Artemia salina*
Other well-known name	:	Artemia, Brine shrimp, Brine worm, Sea Monkey (in USA)

Introduction

Of all the live food used for larvae culture of finfish and shellfish in aqua hatcheries, *Artemia* is the most widely used organism. The main advantage of using *Artemia* is that one can produce live food "on demand" through hydration of dormant cysts into nutritious nauplii within 24 hours.

Identification

Artemia cysts measure about 200 micron in diameter and are brown in colour. It remains covered with hard shell or chorion. Freshly hatched *Artemia* nauplii are about 0.4 mm in length. It is orange in colour due to presence of yolk, containing all essential fatty acids, amino-acids and significant concentration of vitamin, hormones and carotenoids in addition to protein content of nearly 50 per cent. Nauplii have a pair of rudimentary antennae. Adult *Artemia* measures 1.0 to 2.0 cm in length. It is characterized by the head portion bearing a pair of stalked compound eyes, well-developed antennae and the trunk bearing 11 pairs of thoracopods that carry out the function of locomotion, respiration and filter feeding. The elongated abdominal portion extends behind the thoracopods and ends into the anal furca.

Habitat

Artemia is habitats saltpans in different period of the year. This is dependent on salinity conditions. It occurs only in specific areas for uncertain period. It occurs mainly in high salinity water.

Food and Feeding

In nature *Artemia* chiefly feeds on algae. It prefers to take and digest *Dunaliella* and *Scenedesmus* but cannot digest species like *Chlorella* spp. due to thick cell wall. In

Figuer 5.13: Mature Female (Left) and Male (Right) of *Artemia*

laboratory, it takes any particulate food such as bacteria, pig manure, chicken manure, phytoplankton, rice bran, shrimp head and copra meal. Food particles are brought to the mouth by constant movement of its appendages.

Reproduction

Artemia salina attains sexual maturity in two weeks time after hatching and reproduces continuously throughout its life span of 30 to 40 days.

Artemia occurs in two stains, one bisexual (presence of male and female) and the other parthenogenetic (only female). Adult male of *Artemia* can be identified by presence of a pair of large sized claspers in its anterior portion. *See* Figure 5.13 Adult females, on the other hand, are devoid of such claspers and possess an ovisac and uterus.

In case of bisexual strain, the sperm from the male is transferred to the ovisac of the female during the course of copulation, thus, the eggs present in the ovisac are fertilized and further embryonic development takes place. On the other hand, in case of parthenogenetic strain, embryonic development starts directly as soon as the eggs reach the ovisac. The embryo develops either into nauplii or gets coated with shell to form cyst in the ovisac, depending on the prevailing environmental conditions of water medium. Such choice between oviparity (release of eggs) and viviparity (release of nauplii) is common train of *Artemia*. It is observed that low salinity up to 70 ppt with aerobic condition, induces release of nauplii while higher salinity triggers release of eggs.

Sexually copulated or parthenogenetic female when gravid carries about 100 to 110 eggs in the brood sac. Eggs are generally released in 4 batches at an interval of 4–5 days (Reddy, A.K. and Thakur, 1998).

Life Cycle of *Artemia*

The lifecycle of *Artemia* pass through 13 instars (larval stages) as shown in Figure 5.14. Though the first instars (*i.e.* nauplii) is the best as larval feed, nevertheless, its other instars and adults are also used as live feed in various aquaculture operations.

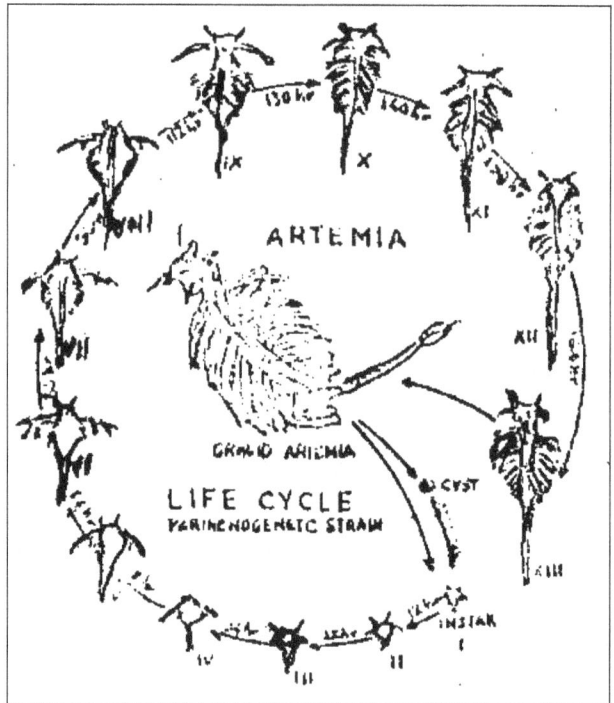

Figure 5.14: Life Cycle of *Artemia*
(*Source*: **Reddy, A.K. and N.H. Thakur, 1998**).

Requirement of Artemic Cyst

To produce one million post larvae of *M. rosenbergii*, the requirement of cyst can be calculated with 25 per cent survival from stage-I larvae to post larval stage works out to about 9000 million nauplii.

Hatching of *Artemia* cyst

Step-I

Put dry cyst in seawater (20 ml of water may be used for every one gram of cyst). After one-hour filter through 100 micron mesh blotting cloth.

Step-II

The hydrated cysts are kept in 5 per cent Sodium hypochloride (NaOCl) solution @ 15 ml for every one gram cyst. The temperature may increased beyond 40° C. To avoid this keep in cold water or put ice cubes out side of the containers.

In about 5 to 10 minutes, Chlorion (Cover of cyst) gets dissolved and decapsulated cyst are filtered on a micron mesh size cloth. Wash thoroughly to remove toxic effect of chlorion. In order to ensure complete removal of chlorion, the decapsulated cysts are given a dip in 0.1 per cent Sodium thiosulphate solution.

Step-III

Decapsulated *Artemia* cysts are to be hatched. The optimum condition of water quality is to be kept as under.

Temperature	:	27 to 30° C
pH	:	7.5 to 8.5
Salinity	:	25 to 30 ppt
Light	:	1000 Lux
Dissolved Oxygen	:	Saturation point.

Step–IV: Harvesting

The freshly hatched *Artemia* nauplii are harvested in 100-micron mesh net by taking advantage of their photo static nature. Put in harvesting Jar. Aeration should be stopped and cover harvesting Jar with thick cloth keep for some time. All unhatched cysts settled in cone. Remove using valve. Put torch or bulb on the transparent side of hatching Jar. All nauplii will be gathered near light. Remove nauplii in the net by opening valve. Keep net in water or in bucket having an out let at the upper edge to facilitate drainage of excess of water.

(17) Cyclops

Classification

Common Name	:	Cyclops
Phylum	:	Arthropoda
Class	:	Crustacea

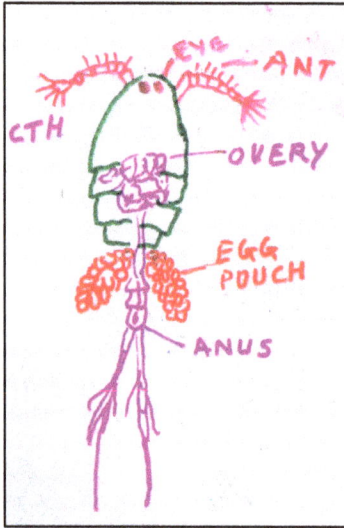

Figure 5.15: Cyclops
CTH = Cephalothorax, ANT: Antinules

Sub Class	:	Copepoda
Order	:	Calanoida
Family	:	Cylopoids

According to references there are 47 species of cylops are found in India. They are found in brackish water, freshwater pond and ditches. They are free living, commercial and parasitic. The body is elongated, somewhat oval in shape. Eggs are kept in abdomen of female in 4 or 2 egg sacs. They are having six pairs of thorasic appendages. Abdomen is without appendages but provided with a pair of caudal style. Antenules and antenna well developed.

Uses

It is an important feed for fry as well as live food for fishes.

(18) *Chironomid*

Classification

Common Name	:	Blood Worm
Phylum	:	Arthropoda
Class	:	Insecta
Order	:	Diptera
Family	:	Chironomidae
Scientific Name	:	*Chironomid*

Introduction

Chironomid larvae are commonly used as live food for maintenance of ornamental fishes.

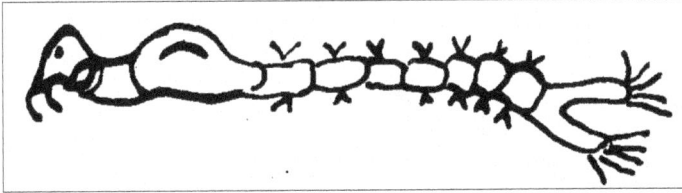

Figure 5.16: Blood Worm

Habitat

Chironomid flies attracts towards foul smell where organic hatch into Chironomid larvae. As larvae grows come out of tubes and swim vertically in water. Fully-grown larvae are dark in colour.

Identification

Chironomid larvae looks like an Annilid worm. Body is segmented. Head is free. It is about 1.0 to 1.5 cm in length. Larvae has three legs, one in front and the other two at rear end of body. Rear end (tail) has hairs.

Food and Feeding

The worm Chironomid larvae are herbivorous in feeding habits and feed on algae, detritus, decaying organic and vegetable matter etc.

Reproduction

Chironomid flies lay eggs on organic matter, which is immerged in water. The eggs hatch straight into proboscis larvae. Female lay a hatch of about 20,000 eggs, which hatch out in about 3 days.

Culture

Chironomid can be cultured in 70 liters tank containing palm oil mill effluent (POME) and *Chlorella vulgaris* separately. The production of Chironomid larvae is significantly higher in POME tanks (58 g. 20 liters POME) than Chlorella culture (35 g./20 liter Chlorella).

Table 5.2: Feeding Schedule for Baby Fishes (Fry)

Size of Fry	Feed for Fry				
	First Stage	Second Stage	Third Stage	Fourth Stage	Fifth Stage
Small (*e.g.* Bettas, Gaurami)	Green water, prepared dry food	infusoria	Large infusorians	Brine shrimp nauplii, small Daphnia	Small dry food, large Daphnia
Medium (*e.g.* Barbs, Characins)	Egg yolk, infusoria (Small)	Small dry food, Rotifers	Brine shrimp nauplii, small cladocerans	Large dry food	Large Daphnia, Chopped worms, large dry food
Large (*e.g.* Live bearers, Cichlids)	Brine shrimp nauplii, small dry food	Small Daphnia	Large Cladocorans	Chopped worms, large dry foods	Chopped worms, large dry food

Source: Alappat and Biju Kumar, 1997.

Chapter 6

Ornamental Fish Breeding Technique

Introduction

One of the attractions of ornamental fishes is that they breed in captivity. The techniques used to facilitate their breeding are however elastic, so much so, efforts in regards to their breeding vary from breeder to breeder.

Breeding of fish means selection of paired breeding fish, reproduction and successful rearing of fry until they reach the juvenile stage. Internal and external both factors play important role in fish breeding. Temperature change of water is one of the most important factors. Different values are found for every species. Other factors are physical and chemical properties of water, along with all these current of water is also important.

Sex Determination

In order to breed a species, the aquarist primary needs to be able to distinguish between sexes. Determination the sex of a fish is an important step to know whether one has a pair or not. The sexes can be easily distinguished by primary (shape of sex organs) and secondary differences, (size, shape, colour, sexual dichromatism fin development). Males are generally more colourful, larger and have more elaborate fins. In some species, the males are slightly larger and the females have a belly slightly rounded.

Selecting the Parent Fish

One males and females have been distinguished, suitable pairs or spawning groups have to be chosen. In some species ratio of male and female are different. There are several important trails to seek in choosing the parent fishes. These consist of:

1. Choosing fish that are healthy and display good feature like strong colouration, good fin development, etc. These are indicative of production of attractive and healthy young ones without deformities.

2. Ensuring that the selected pairs are compatible. If sometimes happens that the brooders placed in a breeding tank do not get along well. In fact, with many cichlids, pairs form only after a group has been raised together for a long duration. In certain species, one partner will bully the other to death if there is no compatibility.

3. It has to be ensured that the paired sexes are of the same species. This is necessary to prevent emergence of hybrids which usually sterile. With some cichlids and Killifish, females of different species look similar.

Conditioning the Fish to Breed

In the wild, breeding instincts is stimulated by a change in the environmental conditions around the fish, and this can be created to some extent in aquaria. Presumably the circumstances that trigger breeding are complex, consisting perhaps of combination of factors such as food availability, water temperature, length of day light, changes in water chemistry etc. A varied diet, with an increased level of protein is recommended for conditioning. Many species can be conditioned using a well-balanced flake food, through conditioning with live foods such as brine shrimp, insect larvae, and flying insects gives better results. A small increase in the ambient temperature can prove to be beneficial to stimulate breeding instinct. Further, more of lighting provides a stimulus for coldwater fish species which are normally exposed to seasonal changes. The condition of the water is significant and shifting a pair to a fresh tank may lead to breeding success. It is desirable that sexes are separated three weeks before being re-introduced. Fishes not indulging in breeding can be induced to do so by injecting them with specific hormones to stimulate reproductive activities. Such techniques, however, are not normally within the reach of the aquarists. (Alagappan, M. and Vijula, K. 2004)

Basic terminologies in breeding are as under:

1. Oviparous

Animals that lay eggs (Figure 6.1).

2. Ovo-viviparous

Animals that produce eggs nutrition other than through yolk in the eggs. Embryos develop within female but are nourished principally by the yolk sac (Figure 6.2).

3. Viviparous

Animals that provide on going nutrition for the developing embryo, which may be achieved in a variety of ways. Embryos develop within female and receive nourishment from their mother (Figure 6.3).

Sexual Dimorphism

Sexual differentiation is characterized by the fact that the male and female

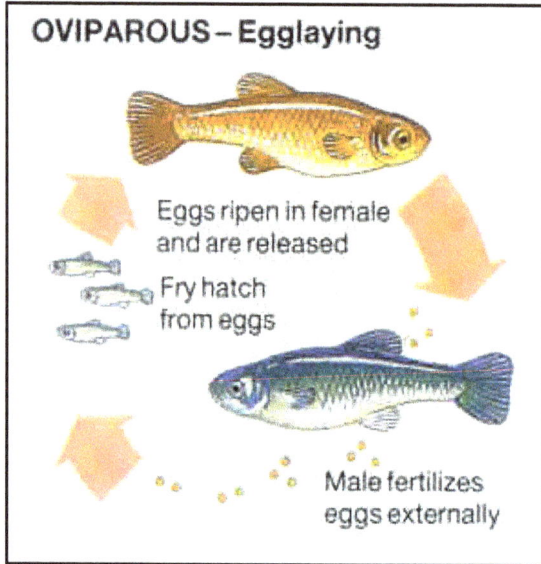

OVIPAROUS – Egglaying

Eggs ripen in female and are released

Fry hatch from eggs

Male fertilizes eggs externally

Figure 6.1

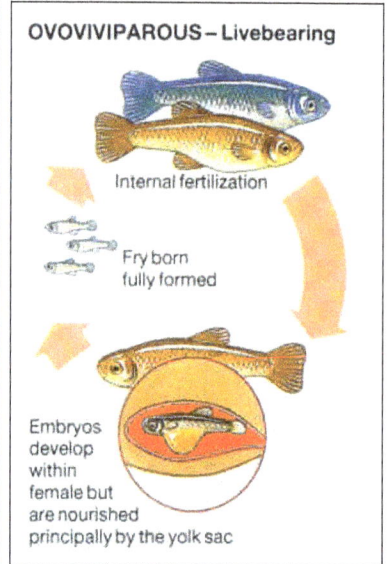

OVOVIVIPAROUS – Livebearing

Internal fertilization

Fry born fully formed

Embryos develop within female but are nourished principally by the yolk sac

Figure 6.2

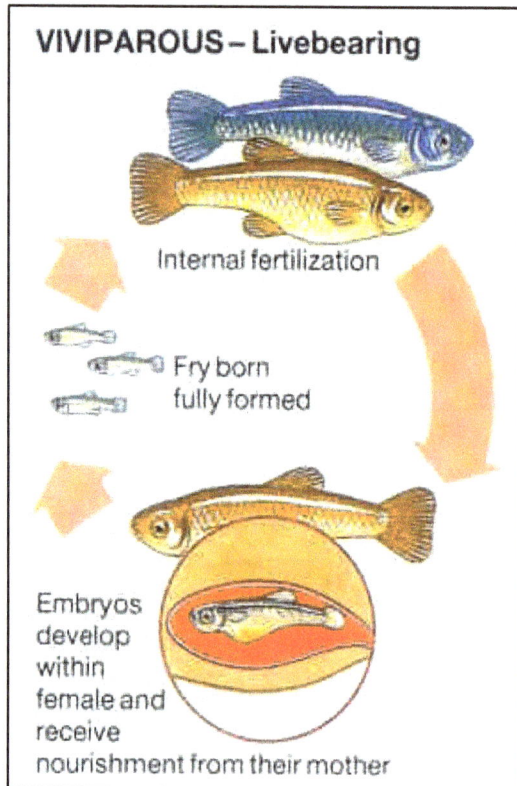

VIVIPAROUS – Livebearing

Internal fertilization

Fry born fully formed

Embryos develop within female and receive nourishment from their mother

Figure 6.3

reproductive glands or primary characteristics of sex, occur separately. In many species of fish the external difference between the male and female are minute or practically non-existent; such fish are said to be mono-morphic. In some species, however, sex characteristics are very pronounced and in such case we speak of sexual dimorphism. Sexual dimorphism in general evidenced in the bright colouration of male (very occasionally of the female).

During the spawning period certain sexual difference become more pronounced, particularly the colouration.

Type of Brooders

Breeding techniques and devices involved in breeding differ according to the spawning behaviours of the fish. The spawning behaviours differs with each family variations are also present among members of the same family. Two major groups can be made depending upon their breeding methods.

    (A) Egg layers

    (B) Live bearers

The egg layers can be classified into different categories depending on their breeding habits (Alappat and Biju Kumar 1997).

    1. Egg scatterers

    2. Egg hangers

    3. Egg depositors

    4. Mouth brooders

    5. Nest builders

    6. Egg barriers

(1) Egg Scatterers

The egg scatterers either spawn in pairs or in groups. The compensatory factor is that large amounts of eggs are produced and because of this some of the eggs have a chance to get fertilized and hatch out. The egg scatters release the egg in a haphazard manner, scattering the eggs in all directions. Once the spawning is over, they often consume the eggs. The egg scatters show no parental care. Hence, it is the job of the aquarist to protect the eggs from being eaten by the parents. The egg scatterers usually show less interest in the eggs while the spawning is in progress. They start feeding on eggs only after spawning is over.

It is, therefore, advisable to use spawning greed (for non-sticky eggs) or Kaka Bandh (for adhesive eggs).

Denios, Gold fish, Glass fish, Kissing Gouramies, Tetras, Barbs, Rasbora etc. are coming under adhesive type eggs layers group. (Figure 6.4) Barbs and Danios, lay non-adhesive eggs (Figure 6.5).

Breeding Techniques

Figure 6.4: For Sticky Eggs Layers

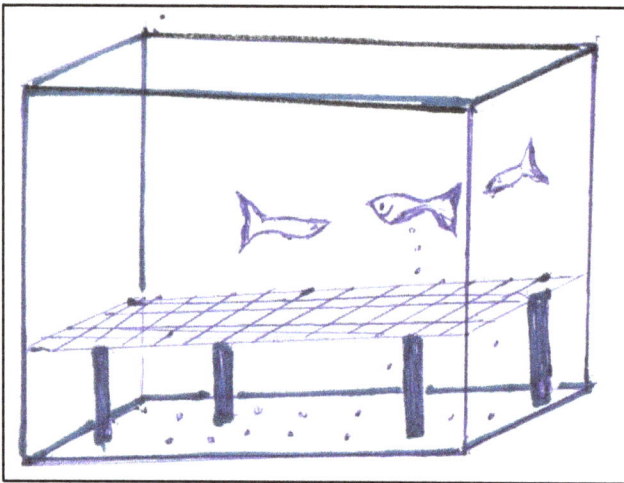

Figure 6.5: For Non-Sticky Eggs Layers

(2) Egg Hangers

Some groups of fishes, particularly Killifishes, attach their eggs to plants or to other hanging objects by means of fine sticky thread. The process of egg laying may last for several days. As a general rule, Killifish do not eat their eggs. However, it is better to remove the eggs as soon as the spawning is over. 2–3 males may be provided for one female for better fertilization. A bunch of threads or plants may be used for collecting eggs (Figure 6.6). The eggs attached to the threads or plants may be kept in another tank for further development. Since the incubation time is high (2–3 weeks), it is necessary to feed the young ones as soon as they emerge from egg, they should be fed with *infusoria* and Artemia nauplii. Fry should be removed to larger tanks.

**Figure 6.6: Technique for
Sticky Eggs**

Killi fishes like *Panchax* spp. are coming under this group.

(3) Egg Depositors

The egg depositor care more for their young than egg–hungers or scatters.

They show parental care. They need substratum like leaves, rocks, flowers, pots, tiles or slates to deposit the eggs, most of the depositor select their own pair/mate. It is best to remove pair to water with similar condition. Transfer the eggs to a tank treated with methylene blue. The eggs are to be given perfuse aeration. The fry once

Figure 6.7: Breeding Arrangements for Cave Lovers

they become free swimming, even through the fish shows parental care, fish should be removed.

Cichlids, like angelfish, convict cichlid etc. are coming in this group. These species do not usually resort to eating their own eggs. Among egg depositors that care for their eggs are chchlids and some catfish. Cyprinidis, and Killifish, constitute the majority of egg–depositors that do not care for their young ones.

Some species such as Discus and Angelfish prefer vertical surface.

For cavity Spawners, follower pots plumed towards their side, coconut shells and rocky caves are suitable spawning sites. The tank should be furnished with wither live or artificial (plastic) plants to give the fish sense of security (Figure 6.7).

(4) Mouth Brooders

Mouth brooders carry their eggs or larvae in their mouth. Mouth brooders consist of ovophiles and larvophiles. Ovophiles or egg living mouth brooders lay their eggs in a pit. These are sucked into its mouth by the female. These eggs, large in size but few in number, hatch in the mother's mouth and the fry remain there for a period of time. Many Cichlids and some labyrish fish are ovophile mouth brooders. Larvophile or larvae loving mouth brooders lay their eggs on a substrate and guard them until the eggs hatch. After hatching female picks up the fry and keeps them in her mouth (Figure 6.8). They are released at a stage when they can feed on their own. (Alappan and Vijula, 2004)

Figure 6.8: Mouth Brooder

Ovophile mouth-brooders can be bred in the main aquarium because the eggs are protected in the mouth cavity. Larvophile mouth-brooders should be placed in a separate breeding tank because the eggs are not protected in the mouth, but laid on a surface where they are exposed to predators.

These mouth brooders usually prepare breeding pits on the bottom sand on which spawning takes place. Hence, the aquarium bottom should be made of sand or fine gravel. In case of Tilapias, the female carries the eggs in her mouth. The eggs hatch inside the mouth and the parent carry the fry also for some time.

(5) Nest Builders

Some fishes like Fighter fish, gouramies and paradise fish build bubble nests

Figure 6.9: Male Fighter Fish taking Care of their Eggs

during the time of breeding. The bubble nest is usually built under floating leaf or on the sides of the aquarium. The bubble nest is made of fine bubble coated with saliva. Aerating the aquarium should be avoided since this result in the destruction of the bubble nest. The male guard the eggs once the spawning is over (Figure 6.9). The female should be removed as soon as spawning is over, while male should be removed as soon as fry become free swimming.

Two compartments should be provided in aquarium, one big and one small. In small section female Fighter fish may be placed. And in bigger section male should be placed (Figure 6.10). Play before breeding starts both male and female dash on glass

Figure 6.10: For Bubble-nest Forming Egg Layers

partition. Male builds a bubble nest by gulping air from the water surface, mixing it with saliva and blowing out the bubble. A number of bubbles are blown and accumulate at the water surface in the form of a "nest". The size of nest is 7 cm in length and centimeter in a height. A healthy female is introduced in to the tank and courtship starts. The male spreading its fins and opening wide its gill covers. If the female is ready for breeding, it responds with spreading fins. Now male leads the female towards the nest. If the female is not ready for breeding, the male will attack her and even killing her. If female is ready for breeding, mating starts. The male drivers the female towards the bubble nest. Under the nest, male folds his body around the female and they both sink together, female release few eggs (Figure 6.11). If these eggs drop down towards the bottom, the male picks up the eggs in its mouth and shoots them into the bubble nest. This process is repeated for one to two hours. 100 to 500 eggs are collected in the nest. At this stage, the female should be removed from the breeding tank.

Figure 6.11: Male of Fighter Fish Squeezes the Female until She Releases the Eggs

(6) Egg Barriers

Fishes that come under this group usually inhabit water that dries up at some part of the year. The majority if egg barriers are Kill-fish, which lay their eggs on a muddy bottom. The parents mature very quickly and lay eggs before the water dries up. They die later after laying eggs. The eggs remain in a dormant stage until rains stimulate hatching.

A peat moss substrate is one of the best substrates for egg burying species. The peat moss can be stored for weeks to months (depending on the species). In order to initiate hatching the stored peat can be immersed in soft water. A new peat moss can also be placed in the tank to encourage further spawning (Figure 6.12).

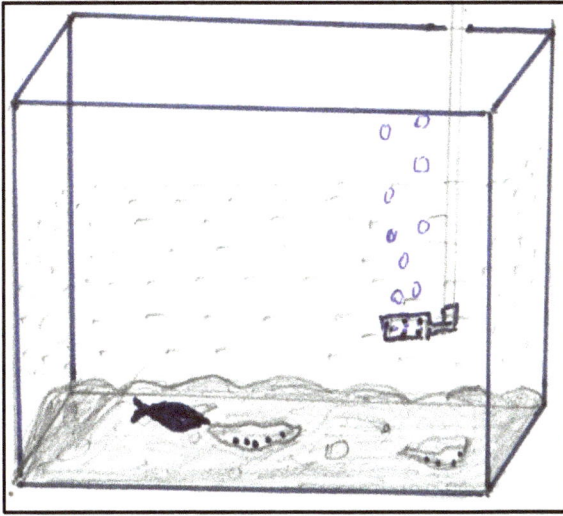

**Figure 6.12: For Pit Loving
Egg Layers**

Many fish deposit their eggs on flat stones or branches near the bottom (Lithophilic fish) (Figure 6.13). These are mostly members of the Cichlidae family, which provides the eggs and fry with intensive care. Their members include also species that take up the eggs in to their mouth (Mouth brooders).

Figure 6.13: For Fish Laying Eggs on Stone

Labyrinth fish generally deposit their eggs in bubble nests on the surface. Some catfish *e.g.* of the Callichthydae family hide the nest under the broad floating leaves of water–lillies under fallen roots etc. (Figure 6.14). The secretion cementing the air bubble together contains bacteriostatic substance which protects the eggs.

Livebearers

A livebearer fish is that employs internal fertilization of eggs followed by either their deposition after a shorter or longer period of time or their retention within the body of the female until the movement of birth (Figure 6.15). There are two types of

Figure 6.14: Sticky Eggs in Nest Under Big Leafed Plant

Figure 6.15: Female Guppy Showing Gravid Spot (Livebearer)

livebearers, Ovoviviparous and Viviparous. In Ovoviviparous fishes the eggs hatched within the female before they are released. In the case of others, the Viviparous ones, the young ones, formed within the body, are released. Livebearers are often prolific breeders and can be bred easily.

The live bearing fishes are the easiest of all aquarium fishes to breed. Guppy, Mosquito fish, Sward tails Mollies etc. are coming under this group.

Raising the Fry of Livebearers

The young of livebearers are somewhat larger in size. So much so, they can feed on dry or other prepared food straight way. If they are given only prepared food, growth will be poor. However, a mixture of live and dry food is consumed well. In the early stages, feeding with a component of live food is very important for good growth. Later, live food matters much less, although the fishes will grow better with a good proportion of live food. Suitable first stage live foods are micro worms, newly hatched mosquito larvae or shredded earthworms, Daphnia, shredded white worms. Suitable dry feed includes any fine powdered food; such as finely grouped dried shrimp, fine cereals, and liver or egg powder (Alagappan and Vijula, K. 2004).

Less than 3 per cent of the 18,000 or so species of bony fishes are live bearers, including marine fishes.

Fish keepers choose live bearers for the aquarium for a number of reasons, some simply want bright coloured fishes which, given suitable conditions. Some prefers to keep the "Cultivated" varieties and pursue new colour variations, while other keeps them for rearing varieties simply for their beauty and function.

Fertilization

External Fertilization

As a rule, when spawning the female release the eggs in the water and the male simultaneously releases milt close to the eggs, the eggs are thus fertilized out side the body of the female; this is called external fertilization.

Internal Fertilization

It is method by which the male can get the milt, containing the sperm, into the body of the female to achieve internal fertilization. Once fertilization has been achieved, the female must make provision for the developing embryo to receive nourishment. This is very important because, in evolving a system of live bearing, most fish have sacrificed the ability to produce large number of eggs in return for the alternative of giving fewer fry a better chance of life.

The problem of internal fertilization has been solved different ways in the various families, generally through a modification of the anal fin to form the equivalent of a penis.

In *Poecillids*, the first rays of the anal fin are modified to form a long pointed gonopodium. This can be swung forward, becoming proved in the process, to allow the passage of packets of sperm, or spermogoeugmnata. In these packets all the sperm have their "tails" embedded in mucus core and their "head" face outwards. One mating can result in up to 3000 of these packets being passed in to the female. These packets break up to release the sperm, some of which fertilize the ripe eggs, while the remainder are stored in the fold of the oviduct wall for fertilization of successive broods.

Right Handed and Left Handed Mating

In this type of mating, the gonopodium bands to one side and male can only mate with a female with genital opening inclined to the opposite side. This is right handed male mate with left handed female. The female has an enlarged scale which obscures the opening from one side or the other, and it is this that causes the left or right handedness.

In connection with internal fertilization, we find the development of ovoviviparity in some groups of fish. Where the fertilized eggs develop within the maternal body until hatching and extruded as living young. Ovoviviparity differs from viviparity in that the mother does not nourish the embryo.

The shape of the eggs is typical for each species of fish and may be round, oval, barrel-shaped, tear shaped etc. The size of eggs and the amount laid depends not only on the species of fish but also on the age and size of female. A smaller number of minute eggs may be expected during the first spawning of young females. As they grow older and bigger up to certain age depending on species they produce eggs better quality and in greater number.

Breeding of Livebearers

Breeding of live bearers is easy with good survival of young ones if little attention is given on the process. Fishes like Guppy, Platy, Molly and Swordtail are live bearers. These are also known to have cannibalistic habit that means the adult may eat their young ones. Thus, the brooders must be kept in wire meshed breeding traps, fixed inside the breeding tank, through which the young ones escape when born (Figure 6.16). Planting with large number of bushy plants also provides safety to the young ones.

Figure 6.16: Breeding Cage of Livebearer

For breeding live bearers, the males may be distinguished by the presence of gonopodium which is used to fertilize the eggs. In case of guppy with each act of fertilization, the female can brood three to eight batches of developing eggs. The females ready to breed are identified with their bulging belly.

Live bearers need slightly acidic water for breeding. The selected male and female are kept in separate tanks for a week, before introducing them in the breeding tanks. Here they are fed with protein rich live food. Generally, the male and female fish are released in the breeding tanks in a ratio of 2: 1. The act of laying young ones is observed periodically and the young ones and adult fish are kept separate using a fine meshed barrier or dense plantation. The young ones are fed with infusoria and artemia nauplii for one week. From the second week onwards, they may be fed with strained out daphnia, Cyclops, dry feed etc.

Breeding Techniques

(A) For Nest Forming Fish

There are some fish like Fighter fish forming nest for keeping their eggs.

For them two compartments should be provided in aquarium, one big and one small (Figure 6.10) In small section female fighter fish should be kept and in bigger section male should be placed. Play before breeding starts both male and female dash on glass partition, when fully excited male starts forming bubble nest. As soon as bubble nest is ready remove glass partition. Male leads female to nest for passing through nest. He press females belly and guide her to pass through the nest. When female passes, belly will be pressed and eggs will be released. Then male enters in the nest and release sperm. These practices continue till spawning is over, remove female immediately. Male will take care of eggs. If some fertilized egg fall on bottom, he will pick up in mouth and place in nest. When juveniles starts coming from nest remove male also.

(B) For Non-sticking Egg Layers

Some fishes which lay non sticky eggs, breeding greed should be provided for such fishes. Breeding greed with net should be kept about 5 to 10 cm above the bottom (Figure 6.5). The size of net should be such that parent fish can not pass through it. Non sticky eggs will fall down from the net so that parents may not eat them. Sperm released by male fertilize these eggs. After release of eggs remove female.

(C) For Sticky Egg Layers

There are many fishes which lay sticky eggs. Different arrangements are required to be organized for such different fishes. Details are given for such fishes in text. Fishes like Angel stick their eggs on sloppy surface (Figure 6.4). Fishes like Gold fish etc stick their eggs on Kaka Bund (Figure 6.6). Some fishes stick their eggs under big leaf of aquatic plant (Figure 6.14). Fishes like Oscar is cleaning and then sticking their eggs on stone or cave (Figure 6.13).

Breeding in Marine Fishes

Marine fishes employ various methods of spawning, in much the same way as their freshwater cousins; these include egg scattering egg depositing, mouth brooding and pouch-brooding.

Breeding Season

Breeding season of various fishes are given in Table 6.1

The egg depositors feature anemone fishes demersal fishes and many gobies, whose spawning behaviours and parental care is similar to that found in freshwater cichlids. The jaw fishes and cardinal fishes incubate eggs in their mouths, while the male seahorse is famed for adopting a similar role with eggs deposited in his abdominal pouch by the female.

**Table 6.1: Data Base Representation of the
Breeding Season (Months) of Varies Fishes**

Sl.No.	Variety	Jan.	Feb.	Mar.	April	May	June	July	Aug.	Sept.	Oct.	Nov.	Dec.
1.	Gold fish	Y	Y	Y				Y	Y	Y			Y
2.	Koi carp	Y	Y				Y	Y	Y				Y
3.	Molly			Y	Y	Y	Y	y	Y	Y	Y		
4.	Platty			Y	Y	Y	Y	Y	Y	Y	Y	Y	Y
5.	Sward tail				Y	Y	Y	Y	Y	Y	Y		
6.	Guppy	Y	Y	Y	Y	Y	Y	Y	Y	Y	Y	Y	
7.	Gourami				Y	Y	Y	Y	Y	Y	Y		
8.	Angel			Y	Y	Y				Y			
9.	Fire mouth cichlid			Y	Y	Y	Y						
10.	Congo tetra				Y	Y	Y	Y	Y	Y	Y	Y	
11.	Diamond tetra				Y	Y	Y	Y	Y	Y	Y	y	
12.	Black wido tetra				Y	Y	Y	Y	Y	Y			
13.	Neon tetra	Y	Y				Y	Y	Y	Y			
14.	Cardinal tetra	Y	Y				Y	Y	Y	y			
15.	Golden wido tetra						Y	Y	Y	Y			
16.	Serpe tetra						Y	Y	Y	Y			
17.	Barbs			Y	Y	Y	Y	Y	Y	Y			
18.	Siamese fighter				Y	Y	Y	Y	Y	Y	Y		
19.	Tiger barb			Y	Y	Y	Y	Y	Y	y			

Y: Breeding.

Source: Lokenath Chakraborty, 2006.

Within the frame work of the different spawning methods, there is also a diversity of breeding behaviour. Fishes may, for instance, from long-term partnerships; a male may set up an attendant harem of females; shoals of fishes may seasonally congregate for mass spawning; or a male and female may spontaneously spawn as and when the opportunity occurs.

Seasonal conditions can also affect spawning activity. Many species are also heavily influenced by the lunar phases, water temperature and photoperiod (length of day light).

When compared with freshwater, their number of marine species successful reproduced in captivity is very small indeed. The reasons for comparative lack of success are easy to understand. Firstly, marine fish keeping is a young hobby, and much of the original work concerned itself with just keeping the animal alive, let alone breeding them.

Chapter 7
Freshwater Ornamental Fishes

In this chapter details of 171 freshwater Ornamental fish is given. Details includes classification, common name, Native of respective fish, Identification, Characters, Sexual dimorphism, Sex ratio for breeding, breeding method – Technique, Water quality at the time of breeding and some times even for newly hatched fry, care to be taken during breeding and culture of fry and adults, feed for fry and adults.

All 171 fresh water fishes are divided in different families (A to Q) and groups. Ornamental fishes covered under different groups are as under Figures in bracket shows the number of fish covered under the same group:

Fighter fish (1), Indian Catfish (1), Betta (2), Corydoras (9), Glassfish (4), Snakeheads (7), Gourami (9), Hatchet fish (2), Tetra (12), Cichlids (17) Discuss (4), Botia/Loach (7), Barbs (8), Danio (4), Gold fish (14), Rainbow fish (4), Rasbora (6), Panchax (4), Eel (2), Badi (3), Molly (7), Guppy (2), Oscar (1), Puffer fish (8). Mouth Brooder (1), Red tailed Black Shark (1) Over and above Photographs of Angel (14), and Guppy (18) is given. Hence details of 171 fishes are given.

Species wise details of above fishes are given here under.

(A) Siamese Fighter Fish

Common Name	:	Siamese Fighter Fish

Classification

Phylum	:	Chordata
Super class	:	Pisces
Class	:	Osteichthyes
Order	:	Perciformes

Sub order	:	Anabantoidei
Family	:	Anabantidae
Scientific Name	:	*Betta splendens* (Regan 1909)
Other Species	:	*Betta trfasciata, Betta pungnax*

Fighter Fish (Male)   **Fighter Fish (Female)**

Figure 7.A.1

Introduction

Siamese fighting fish is native and found in Indonesia and Malaysia. This fish is very well known in India. The attractive colour and hardness of the species are the characters of the species for its wide adoption by the hobbyists. Their growth is rapid.

Identification

The dorsal and anal fins are furnished with spines. The male fighters are the most beautiful. Adults develop a most spectacular fin age a high dorsal, large flowing tail and a deep anal fin. Another characteristic is the long and pointed fins which are thrust forward when the fish is excited. When it is in fighting mood, placing upward stimulates his anal fin and it holds all fins stuffy erected. The fins (except the ventral fins) are longer than those of the female and under certain conditions are capable of attaining extreme length and shape.

Colour

"Corn flower" blue and fairly red are the two main colours. But there are many colours variations including albumin, dark blue, green and red all with metallic quality the colour completely converting the whole of the body and fins. Females are less colourful and do not possess the layer flowing fanners as of the males.

Sexual Maturity

Some species attain sexual maturity at the age of five weeks, but the best breeding results are achieved with a brooder five or six months old. The ability of the male's fins to grow to various lengths has been put to good use by breeders. When they are sexually mature the males are put in separate place one beside the other.

Sex Ratio

1 male : 1 female

Water Quality for Breeding

Temperature	:	26–30° C
pH	:	7.0
Hardness	:	Max. 2° dCH

Water quality is not critical but breeding appears to be stimulating by keeping the water level low; such as 15 cm (6 inch) or less.

Breeding

Male builds bubble nest usually under or adjacent to floating leaf. It is, therefore, necessary to keep floating leaf in the aquarium. It was found that the bubbles contain bacteriostatic substances and substances that have a beneficial effects on the chemical composition of the water in the vicinity of the eggs. When a male starts building a nest, or even a few bubbles, it indicates that it is ready to breed. Initially male and female are kept separately by keeping a glass curtain in the aquarium. When both are excited and male starts bubbling glass curtain should be removed. Male proceeds to court the female with height end splash colour towards the nest and if all goes well, the male will wrap himself around her and fertilize the eggs as they are released. The male will collect the eggs and plant them in nest. Once the spawning is over, the female should be removed. The male will guard the nest.

Aquarium Care

Carefully take out most of the plants and lower the water level to 5 cm. The intensive metabolism of the growing young fish necessitates frequently cleaning of the tank and regular water replacement. A single male with one or two females can be kept ion a community tank or single species tank of 40-lit capacity or more. However, two males will fight with each other and try to damage each other.

Eggs

About 8 mm size. They sink to the bottom.

Feeding to Fry

Paramecium caudatum, extremely fine nauplii of brine shrimp to cyclops.

Hatching

Fry hatch in 2 to 3 days. Once they become free swimming the male must be removed. Otherwise he will forget all the care and attention and eat the lot.

(B) Catfish

Common Name	:	Indian Glass catfish
Scientific Name	:	*Cryptopterus bicirrhis*
Synonyms	:	*Cryprerichthys bicirrhis, Cryptopterus amboinensis, Silurus bicirrhis, S. palembangensis*

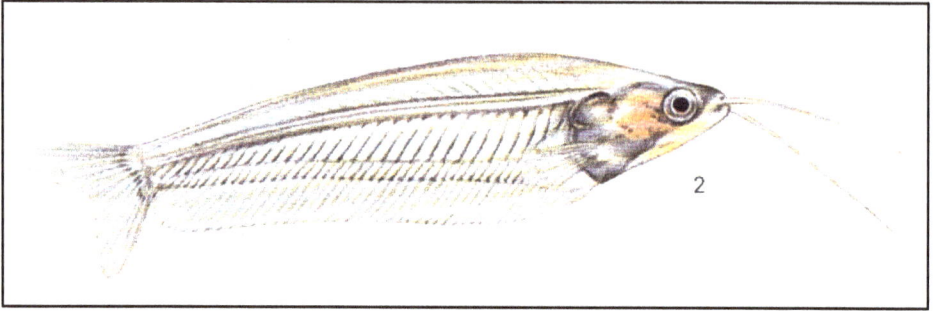

Figure 7.B.1: *Cryptopterus bicirrhis*

Distribution

They are native of Southeast Asia from Thailand to the Greater Sunda Islands.

Identification

They have completely naked skin and two pairs of long barbells. The dorsal fin is either very small or entirely absent. The anal fin is very prolonged and often connects with the caudal fin. There is no adipose fin. The eyes are small. They are exclusively freshwater fish. They can be kept in community tank with small and peaceful species of fish. They are peaceable school fish.

Full Length

Male: 10 cm; Female: 15 cm

Diet

Small live foods may be augmented by artificial foods.

Breeding

Brood fish should be kept in a mono species tank. For breeding purpose set up the fish in a larger shoal. They should be kept in 100–200 liter capacity tank with light and dark bottom. Plant thickets here and there and floating plants should be arranged. Water should be kept streaming with forced circulation filter.

Breeding water temperature 25–28°C, pH 7.0 dCH max.2° may be kept. The lowest temperature limit of the water is 20°C.

Feeding to Fry

Paramecium caudatum, extremely fine nauplii of brine shrimp or cyclops.

(C) Big Mouth Brooding Betta

Common Name	:	Big Mouth Brooding Betta

Classification

Family	:	Belonitiidae
Scientific Name	:	*Betta pugnax* (Cantor 1850)
Synonyms	:	*Betta brederi, Macropodus pagnax*

Identification

The dorsal and anal fins are furnished with spines. The males of many species are territorial and aggressive. The male is more intensely coloured and has longer fin.

Full Length

Adult fish grows up to 10 cm.

Water Quality

Temperature	:	26–28° C
pH	:	6.5–7.0
dCH	:	< 2°

Sex Ratio

1 male : 1 female.

Breeding

During the separate acts of mating the eggs fall in to the arched anal fin of the male, are taken from there by the female, and spat out again in front of the male immediately after. This is repeated a number of times until the male has all the eggs (approximately 50) in his mouth.

Egg

Incubation period in the mouth of the male 12–26 days; the development of embryos is irregular. As the fry successively fill their gas bladders they are left free swimming out side the male's mouth.

As soon as the fry leave the male's mouth they can fully self-sufficient and his task is completed.

Feeding to Fry

Brine shrimp nauplii, adult feed live food.

Indian Paradise Fish

Common Name	:	Indian Paradise Fish also known as Spike Tailed Paradise

Classification

Family	:	Belontiidae
Scientific Name	:	*Macropdodus cupanus cupanus* (Cuvier and Valenciennes, 1831)

Habitat

This fish in habitats pools, rice fields and other stagnant water bodies of India and Sri Lanka.

Identification

This fish not brightly coloured. The body is earthy brown in colour with an indistinct horizontal dark stripe which ends in a blotch at the caudal peduncle, which is prominent in juveniles. The iris of the eye is deep red in colour. The male is darker in colour, especially during the breeding period. The fins are longer and pointed in the males.

Full Length

They grow up to 8 cm length.

Water Quality

Water with pH 6.5 and temperature around 26°C is preferred.

Feed

The fish feed mainly on mosquito larvae in the wild. In the aquarium this can be supplemented with meat bits and dry aquarium feed.

Breeding

In aquarium, always provide hiding places like overturned flower pots, jars etc. Since the fish usually spawn inside them. In their absence the male builds a bubble nest beneath floating plants or leaves. Rest of the breeding and larvae rearing techniques are the same as that in Betta.

(D) *Corydoras*

Classification

Family	:	Callichthydae
Common Name	:	Catfish
Native	:	South America particularly Brazil, Uruguay, Northern Argentina, Venezuela, Peru, Colombia.
Species	:	More than 150 species have been described so far.

Important Species

1. *Corydoras schwartzi* (Figure 7.D.1)
2. *Corydoras ambiacus* (Figure 7.D.2)
3. *C. agassizii* (Figure 7.D.3)
4. *C. leucomelas* (Figure 7.D.4)
5. *C. pulcher*

Identification

The fish has a convex dorsal body profile and flat belly. The colouration and pattern vary depending on the species. The narrow mouth has a pair of stiff barbells. The first rays of the pectoral and dorsal fins are modified in to defensive spines. They

Figure 7.D.1: *Corydoras schwartzi*

Figure 7.D.2: *Corydoras ambiacus*

Figure 7.D.3: *Corydoras agassizii*

Figure 7.D.4: *Corydoras leucomelas*

come in all shapes and sizes with body length varying from 3 to 12 cm, according to species.

C. schwartzi (Figure 7.D.1)

Dorsal spine is creamy white in colour is one of those species that show a wide variation of colour patterns within a single population. It has blotches on the caudal fin.

Corydoras ambiacus (Figure 7.D.2)

It has a black area that starts below the first three or four dorsal fin rays and extends upwards about two third of the way in to the dorsal fin.

C. agassizii (Figure 7.D.3) and C. ambiacus

They both are look similar in appearance at the first instance. They differ by the presence of a blotch on the dorsal fin. The former has dark brown to black pigmentation extending from the body covering the full length of the first two dorsal fin rays.

C. leucomelas (Figure 7.D.4)

It is smaller and its colour is more defined. They have white bodies covered with small jet-black blotches.

C. *paralleus* and C. *schwarti* look alike with body markings in the same fashion. The latter has blotches forming broken bands rarely fused to form solid bands.

Sex

Corydoras are delicate and sexes can be differentiated with broaden body structure; the females are generally stronger than males.

Water Quality Requirement

Type of Water	:	Neutral water
Temperature	:	18–26° C, but ideally 25° C
pH	:	Around 7
Hardness	:	Average

Habitat

They prefer sandy bottom and dense vegetation with many hiding places. They are active and excavate the sand bottom at times. They prefer to be in groups and hence should be kept in groups of 5–8 individuals.

Breeding

They are egg layers. Breeding is somewhat difficult. The ideal temperature for breeding is between 24–25° C. The sexual courtship is simple. The male continues to touch the female from the sides and, if the female is receptive, it will release eggs. The female carries the eggs in a cavity located in the pelvic fins and sticks them in various places. Usually eggs are attached approximately 8–10 cm. below water surface.

Hatching

Eggs will hatch in about 2–3 days and juveniles will start eating micro-organisms after utilizing their yolk sac.

Ideal Aquarium Condition

A well planted tank with many hiding places and appropriate shade. The bottom substance could be coarse and soft sand so that the barbells are not injured while excavating. They prefer a low water level and the water should be clear. They are peaceful and can be kept in community tanks as they go well with other fishes.

Feed

They need to be fed with sinking food, preferably blood worms and brine shrimp.

Stripped Cat Fish

Common Name	:	Stripped cat fish
Family	:	Callichthydae
Scientific Name	:	*Synodontis* spp.
Origin	:	Africa

Introduction

They are very popular and rewarding group of catfish, hardy, peaceful, colourful and amusing. There are more than 80 species of *Synodontis* known. Some of the important species are *S. acanthomias, S. brichardi, S. flavitaeniatus, S. decorus, S. alberti, S. nigrita, S. notatus, S. ocellifer, S. soloni* etc. Once they were regarded as an expensive fish, although it is now a common fish in the hobby. *Synodontis* becomes more in demand as a fish that is compatible with them. They are territorial and will defend their territories.

There are more than 80 species known today in the industry of which following are popular.

1) *Synodontis acanthomis,* 2) *S. abberti,* 3) *S. angelicus,* 4) *S. decorus,* 5) *S. flavitalniatus,* 6) *S. nagiventris,* 7) *S. multipunctatus,* 8) *S. nigrita,* 9) *S. srichardi,* 10) *S. notatus,* 11) *S. ocellifer,* 12) *S. rosertsi,* 13) *S. choutedeni* and 14) *S. soloni.* Details of some of them is given here.

(1) *Synodontis brichardi*

    Scientific Name   :   *Synodontis brichardi*

Identification

It is very distinctive *Synodontis* catfish with long thin built and a mouth more like that of Plecostomus. It is known for its joint teeth or closely spaced lower jaw teeth.

Length

Maximum it grows to 15 cm.

Food and Feeding

They are algae eaters. They eat insect in the wild, while they will accept flake, tablet and live foods in aquarium conditions.

Water Quality

They require good water quality. Ideal water conditions are

    Temperature   :   26°C

    Hardness   :   Neutral to medium hard water conditions.

Aquarium Behaviour

They are territorial but not aggressive. They can be kept with fishes like angels, Gauramies.

(2) Golden Striped Catfish

    Common Name   :   Golden striped catfish or Orange striped squeaker

    Scientific Name   :   *Synodontis flautaeniatus*

Identification

This fish is having orange golden strip from mouth to tail. It is known for its joint teeth or closely spaced lower jaw teeth. It is very popular fish in the hobby.

Length

It can reach an approximate size of 18–22 cm.

Water Quality

Temperature	:	22–25° C
pH	:	6–8

Food and Feeding

They are algae eaters. They eat insect in the wild, while they will accept flake, tablet and live foods in aquarium conditions.

Aquarium Behaviour

It is easy to accommodate and mixes well with Cichlids and robust *anabantioids*. The fish should have plenty of swimming places. Enough hiding places are to be provided in the aquarium.

Breeding

It can be easily bred in captivity.

(3) Clown Syno

Common Name	:	Clown Syno
Scientific Name	:	*Synodontis decorus*

Identification

Juveniles have a long filament on the leading dorsal fin ray, with at an background colouration with black spots over the body. This is one of only three *Synodontis* species with filaments on both maxillary barbells and mandibular barbells. The male has some what ridged genital papillae on which the spermato duct is on the rear side, facing the tail fin. Gravid females will also show extended papillae but the oviduct is on the ventral side of the papillae.

Length

It can grow about 30 cm in aquarium conditions.

Water Quality

Temperature	:	24–28°C
pH	:	6–8

Food and feeding

They are algae eaters. They eat insect in the wild, while they will accept flake, tablet and live foods in aquarium conditions.

Aquarium Behaviour

They are compatible with any small fish but must be kept away from, fin nippers as they may be tempted at the filaments on the dorsal fin.

(4) *Synodontis acanthomias*

Scientific Name	:	*Synodontis acanthomias*

This is one of the largest species of *synodontis* and may be one of the more aggressive ones. With the spotted pattern over the body, it is more desirable than any other larger *synodontis*.

Max. Size

24 cms

Water Condition

Temperature	:	26° C
Hardness	:	Neutral to medium hard water.

Food and Feeding

They eat crustaceans, algae and insect larvae in the wild, while they will accept flake, tablet and live foods in aquarium conditions.

(5) *Synodontis nigriventris*

Scientific Name	:	*Synodontis nigriventris*

Identification

The upside down catfish is an ideal catfish known for smaller aquariums. It is known for its joined teeth or closely spaced lower jaw teeth. These are very popular and rewarding group of catfish colourful and amusing.

Water Quality

Temperature	:	22–28°C. But for spawning 26–28°C
pH	:	7 to 7.5. But for spawning 6.8

Aquarium Behaviour

It prefers over hanging structures or tunnels of rock, wood or large broad–leafed plants in aquarium conditions, which act as shady places for rest while the fish is upside down.

(E) Glass Fish

Common Name	:	Glass Fish
Family	:	Chandidae
Native	:	India, Pakistan, Nepal, Bangladesh, Myanmar and Thailand

Important Species

1. *Chanda ranga* (Figure 7.E.1)
2. *C. lala* (Figure 7.E.2)
3. *C. nama*
4. *C. baculis* (Figure 7.E.3)

1) *Chanda ranga* (Figure 7.E.1)

They are popularly known as Indian Glassfish or Indian glass perch. They are having laterally compressed and some what oval-shaped body. The forehead of the fish is indented; the body is arched. They are having two separate dorsal fins in addition to long anal fin. In reflected light, the transparent body has amber to green iridescence. The fins are transparent. The body colour depends on the

Figure 7.E.1: *Chanda ranga*

area where the fish is found. They exhibit best colours under ultraviolet lights or in moderately lighted aquariums with dark substrates. They are found in still and running waters, clear streams, canals, ponds, and in undated paddy fields. Males have more yellow colour and an iridescent blue fringe on their dorsal fin during the spawning season.

Behavior

They prefer to be in groups. They tend to nip the fins of other fishes especially if the school is not large enough.

Full Length

The maximum attainable size is 9 cm.

Water Quality

Temperature	:	28–30° C
pH	:	7.0–8.5

If 1.0 to 1.5 per cent salt is added it is ideal water condition.

Food and Feeding Habit

In the wide they are carnivorous and thus require meaty food under aquarium condition.

2) *Chanda lala* (Figure 7.E.2)

They are better known as the Indian glass fish.

Habitat

They are found in estuaries areas of India, Thailand and Myanmar

Identification

The body of the fish is with red block spot on the scales. The spinal cord can be seen running through the middle of the body. The female in contrast to the male does not have dorsal fins. These fish occur in large schools in the wild; hence they should preferably be kept in groups in the aquarium.

Figure 7.E.2: *Chanda lala*

Water Quality

Temperature	:	24 to 26° C
pH	:	Above neutral *i.e.* 7.2 to 7.5
Hardness	:	They prefer moderately hard water

Food and Feeding

They ignore most flake and pelleted food and prefer live feed.

Breeding

They could be bred in well planted aquariums provided the water conditions are good.

3) *C. nama* (Figure 7.E.3)

It is known as the elongate glass perch let or painted glassfish.

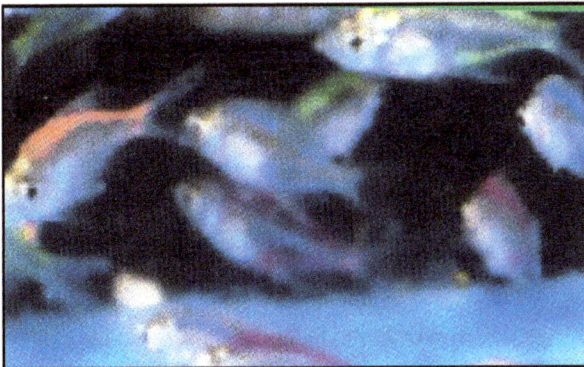

Figure 7.E.3: Painted Glass Fish

Habitat

They are found from the estuaries of the Indian subcontinents.

Identification

The body of the fish is small. This being schooling fish, they prefer to be in groups. Males have slightly longer fins and brighter colours. Most of this species in the trade are painted with fluorescent dyes. They are often injected with non-toxic dyes for the trade to create brilliant neon colours of which fade in about 3–4 months. The dying method results in the stress to the fish, making them susceptible to disease and infection, ultimately causing high mortality.

A recent survey in England indicates that over 40 per cent of painted glassfish were infected with the *lymphocystic* virus, commonly known as cotton fungus.

Water Quality

Temperature	:	26–28° C. However, they can adjust to much cooler or warmer environment.
pH	:	Slightly above neutral 7.2 to 7.5
Hardness	:	They prefer slightly saline water.

Breeding

They can be bred in captivity.

4) *C. baculis* (Figure 7.E.4)

They are commonly referred to as the Burmese glassfish.

Identification

The body of the fish is small with red / black spots on the scales. The spinal cord can

Figure 7.E.4: *C. baculis*

be seen running through the middle of the body. This is the smallest among the glassfish. They are peaceful but shy and prefer sunny locations. They are middle level swimmers.

Maximum Size

The maximum size is 4 cm.

Water Quality

Temperature	:	24–27° C
pH	:	7.0
Hardness	:	They require slightly saline water.

Food and Feeding

They are carnivorous. It would be difficult for them to take flake foods.

(F) Snakeheads

Common Name	:	Snakeheads
Family	:	Channidae
Native	:	India, Asia, Africa, Sri Lanka, Southeastern Iran, Pakistan, Southern China.

Important Species

They are about 28 species of this *Channa* known today. However, following are well known in the aquarium trade.

1. *Channa bleheri*   2. *C. burmania*

3. *C. gachua*   4. *C. micropeltes*

5. *C. Lucius*   6. *C. orientalis*

1) *Channa bleheri* (Figure 7.F.1)

This is widely known as the rainbow snakehead because of its body colouration. This beautiful fish is shaped like a large cigar with fins, the ventral fins having a pattern with the dorsal edged in a brilliant orange. The males are larger than female.

Habitat

It originates from the upper region of northern India, particularly in the Brahmputra river basins of Assam.

Environment

Under aquarium conditions the fish needs plenty of space to move around with plants and hiding places. They need access to the surface to get air.

Water Quality

Temperature	:	22–28° C
pH	:	6 to 7.5

Food and Feeding

They accept live, frozen and also meaty food.

Figure 7.F.1: Rainbow Snakehead *Channa bleheri*

Breeding

They can be bred in aquariums. The eggs float and are guarded by both the parents.

Feeding to Fry

The fry feed on mucus secreted by the parents. Though the fry do not require the mucus to survive, it is said to promote faster growth.

2) *Channa burmanica*

They are popularly known as the Burmese snakehead, originates from the Putao Plains of northern Myanmar.

They are very closely resemble *C. bleheri* in all aspects.

3) *Channa lucius* (Figure 7.F.2)

Identification

They are known as forest snakehead. It has a distinct series of porthole marking on the side and has a more tapering head compared to other species.

Figure 7.F.2: Snakehead *Channa lucius*

Habitat

This species is considered to be endangered among snakeheads. It is found in forest streams and peat swamps.

All other parameters are same as *C.bleheri*. The juvenile fish appear to very different from the adult in colouration.

4) *Channa gachua* (Figure 7.F.3)

Identification

They are known as dwarf snakehead. It is said to be the most suitable *Channa* species for aquarium and one of the smallest and prettiest among the snakeheads. It has

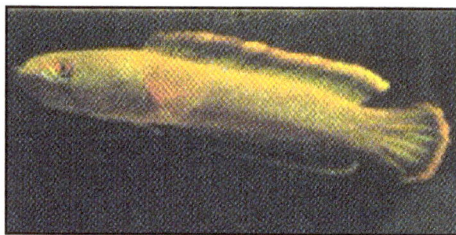

Figure 7.F.3: Snakehead *Channa gachua*

characteristically reddish dorsal and annal fins and distinctly marked pectoral fin.

Habitat

They are found in rainforest streams of Sri Lanka. They also found in southeastern Iran, Pakistan, Bangladesh and Southern China.

Environment

They require relatively well planted larger aquarium with open areas and hiding places.

Water Quality

Temperature	:	24–28°C
pH	:	6.0 to 7.5

Food and Feeding

They accept meaty food

Breeding

This is one of the easiest snakehead species to bread in aquarium. They are mouth brooders.

5) *Channa micropelles* (Figure 7.F.4)

This is also known as giant snakehead or red snakehead. This is the largest of all snakeheads and is cultured commercially as a food fish. They originate from Asia in India, Thailand Myanmar and Malaysia.

They are desirable for aquarium, particularly since young ones display vivid colouration with red being the most dominant.

Figure 7.F.4: Snakehead *Channa micropelles*

Water Quality

Temperature	:	26–27°C
Hardness and pH	:	Values are not critical.

6) *Channa orintalis* (Figure 7.F.5)

They are better known as smooth breasted snakehead. It is native of South Western, Shri Lanka. They are found in its natural habitat ranging from clean freshwater pools to streams. It is a very colourful fish. The colouration of the dorsal fin is a pale olive displaying in the middle, a bluish green band. The outer margin is orange and the tips of the rays, light in colour, body bands cross in the letter part. The round caudal fin is pale yellow in colour with seven dark transverse strips which would be luminous blue some times. The orange colour fins have 4–7 transverse strips. The upper body is light brownish with 10–13 dark bars across the back. The mid lower section is olive brown ventrally the belly and throat colouration being bluish green. There are also

Figure 7.F.5: Snakehead *Channa orintalis*

dark W-shaped transverse band in the flanks. The head has dark lateral line from the snout, through the eye, to the opercle. Several other lines cross the crown of the head. Occasionally a series of minute black spots are seen on the sides.

Water Quality

Temperature	:	24–26° C
pH	:	6.3 to 6.5

Food and Feeding

They can take worms and small fishes under aquarium conditions.

Breeding

They can be bred in captivity. The eggs float and are orally incubated by males

Aquarium Behaviour

It is best to keep in community tank.

(G) Pearl Gourami

Common Name	:	Pearl Gourami

Classification

Phylum	:	Chordata
Super class	:	Pisces
Class	:	Osteichthyes
Sub class	:	Actinopterygii
Order	:	Perciformes
Family	:	Characidae
Scientific Name	:	*Trichogaster leeri* (Bleeker, 1852)

Identification

A dark horizontal line running from snout to caudal fin crosses the lace or mosaic pattern on body. Male possesses long trailing dorsal fin, which reaches up to the caudal fin.

Figure 7.G.1: Pearl Gourami

Full Length	:	11 cm
Sex ratio	:	1: 1

Breeding Water Requirement

Temperature	:	25–28 °C
pH	:	7.0
Eggs	:	They are rich in oil-rise towards the surface
Eggs incubation period	:	24 to 36 hrs.

Breeding Method

The male builds a small bubble nest under the leaf of an aquatic plant in the middle layers of the water. After spawning is over the female and when fry become free-swimming the male also to be removed. The bubble nest is a form of brood protection, ensuring that the eggs and young fry are kept in the upper, oxygen rich layers of the water. Over vigorous aeration and turbulences will destroy a bubble nest.

During this time, it is vital to keep the water shallow less than 15 cm (6 inch) deep and to keep the tank covered; at about two or four weeks of age, the labyrinth organ will begin to develop and any chilling of the air just above the water surface can have disastrous effects on the tiny fry. Similarly, in order for labyrinth to develop properly, the fry must have formation of an oily film on the water surface otherwise that might interfere with the proper development of the labyrinth organ.

Feeding to fry

First 4–5 days *Paramecium caudatum, rotatories*, the *Cyclops* or the finest newly hatched brine shrimp nauplii may be given. At about four to six weeks of age the fry should be feeding on crumble flake foods and similar foods.

Kissing Gourami (Figure 7.G.2)

Classification

Phylum	:	Chordata
Super Class	:	Pisces
Class	:	Osteichthyes
Sub Class	:	Actinopterygii
Order	:	Perciformes
Family	:	Osphronimidae
Scientific Name	:	*Helostoma femmincki* (Cuvier and Valenciennes 1831)
Native	:	Thailand

Synonyms

1. *Holodtoma oligacanthum*

Figure 7.G.2: Kissing Gourami

2. *H. aervus*
3. *H. tambakhan*
4. *Helostoma rudolfi*

Identification

The shape of mouth is unusual when eating or sucking debris from the sides of the tank. These are excellent algae eater. Sometimes two kissing Gaurami comes together and kiss each other. Hence their name is given as Kissing Gaurami.

Full length	:	30 cm
Nature	:	They are peaceful fish. They can be kept in community tank.

Sexual Dimorphism

Indivisible; the posterior end of the dorsal and anal fins with branching rays is sharp cornered in the male and rounded in the female. The female is broader in back. They are egg layers.

Sex Ratio	:	1 male : 1 female

Breeding Water Requirement

Temperature	:	24–26° C
pH	:	7.0
dCH	:	Max 2°
Type of eggs	:	The eggs are very sticky, rise to the surface.
Incubation period	:	50 hours.

Breeding

The female is the more active partner in spawning. Spawning takes in the bottom layers or on float on the surface. The parent fish do not care for the eggs and fry and should therefore be removed after spawning is over. The fry switch over to exogenous nutrition on the fifth day.

Feeding to Fry

Paramecium caudatum, fine artificial fry foods, later brine shrimp nauplii. Besides fish also feed on nano-plankton (the finest phyto planktonic organisms).

Dwarf Gourami (Figure 7.G.3)

Common Name	:	Dwarf Gourami
Family	:	Belantidae
Scientific Name	:	*Colis lalia* (Hamilton Buchanan 1892)
Native	:	India

Identification

It is popular tropical aquarium fish because of small, bright colour, shy but active and peaceful habit. Males have brilliant blue and red alternating diagonal bar but female is silvery with pale vertical bars. They are laterally compressed with an oval-shaped body. The dorsal and anal fins are more developed in male. The pair of ventral fins is filamentous and almost longer than the body. Today more than 10 varieties are cultured such as Golden, Neon, Rainbow, Red Neon, Blue Coral, Long finned, Coral and Peacock. The original Dwarf Gourami bred from spontaneous mutants or developed through hybridization.

Aquarium Behaviour

They are peaceful and hence suitable for community tank. Quite tolerant to different water conditions, it is hardy owing to its capacity for air breathing at the surface of water.

Figure 7.G.3: Red Dwarf Gourami

Aquarium Preference

Fish prefers clam water with groups of plants.

Food and Feeding

The fish is omnivorous, accepting a wide variety of animal and plant food, both manufactured and live.

Breeding Method

Breeding is like Pearl Gourami. Male is preparing bubble nest.

Sexual Dimorphism

They exhibit sexual dimorphism.

Three Spot Gourami (Figure 7.G.4)

Common Name : Three Spot Gourami

Scientific Name : *Trichogaster trichopterus trichopterus*

Other species of the Genus

1. Blue gourami (*T. trichpterous sumaranus*)
2. Pearl gourami (*T. leeri*)
3. Snakeshin gourami (*T. pectoralis*)
4. Moon light gourami (*T. microlepis*)

Identification

The common name is derived from two prominent black spot on the body, the eye is generally considered as the third spot. The common colouration in silvery, olive, darker dorsal and peer towards the abdomen. The dorsal, caudal and anal fins are laced with white or pale orange spots. The long anal fin has pale orange frills.

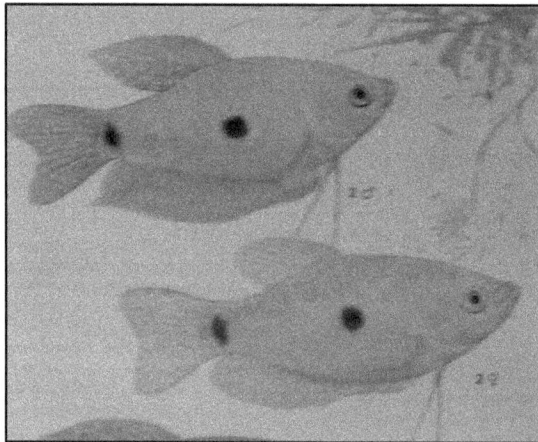

Figure 7.G.4: Three Spot Gourami

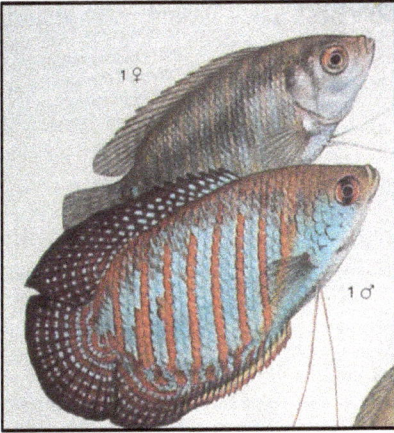

Figure 7.G.5: Blue Gourami (M&F)

The males are slimmer with longer pointed dorsal fins, while females have rounded dorsal fins and plump body.

Breeding

They are prolific breeders and are brightly coloured during spawning.

Chocolate Gourami (Figure 7.G.6)

Common Name	:	Chocolate Gourami
Family	:	Belontiidae
Scientific Name	:	*Sphaerochthys osphoromoenoides* (Canestrini 1860)
Synonyms	:	*Osphromenus malayanus; O. nonetus.*

Figure 7.G.6: Chocolate Gourami (*Sphaerochthys osphoromoenoides*)

Identification

It is a delicate species and does need rather special care in the aquarium. The dorsal and anal fins are furnished with spines. The male of this species is territorial and aggressive. The male often builds a bubble nest. They should be kept in mono species aquarium.

Full Length

5–6 cm in length.

Sexual Dimorphism

Very slight. The male is slimmer. His anal and caudal fins are edged with soft yellow. Mature female tends to be smaller than males and more rounded in over all shape.

Sex Ratio

1 male : 1 female

Food and Feeding

This fish usually accept live food readily although with practice it may be possible to persuade it to accept dried foods.

Water Quality for Breeding

Temperature	:	28–30°C
pH	:	6–6.5 (Slightly acidic)
Hardness	:	10 dCH water should be filtered through biological filter or any other good filter.
Eggs	:	Gourami release about 80–150 eggs with whitish opaque colour of about 1,5 mm size.

Breeding

Female collects fertilized eggs into her well-developed throat pouch and broods them inside for about three weeks. About three weeks after spawning, the fully formed fry 6–7 mm long leave the female's mouth. In order to avoid the fry being eaten by other fish in the tank, remove the brooding female to a small rearing tank about 10 liter capacity containing same water as set up tank and some floating feathered leaved plants. After release of her young transfer her back in to the set-up tank.

Feeding to Fry

The young fry will initially hang motionless among the plants and can be fed on live food or newly hatched brine shrimp nauplii with regular but small feeds and regular partial water changes the fry appears to grow quite rapidly and may reach 14 mm at five or six weeks of age.

Honey Gourami (Figure 7.G.7)

Common Name	:	Honey Gourami
Scientific Name	:	*Colis chuna*

Figure 7.G.7: Honey Gourami

Identification

The fish with typical stocky and compressed body is popular in some markets. The colour is yellowish with a dark brown longitudinal stripe running from the eye to the tail.

Sexual Dimorphism

Sexing is different except during the spawning period when the male shows a distinct colouration while the female has swollen abdomen when she is gravid.

Breeding

Breeding method is same like other Gourami.

Indian Gourami (Figure 7.G.8)

Common Name	:	Indian Gourami
Family	:	Charadiae
Scientific Name	:	*Clisa fasciata*
Native	:	South and Southeast Asia.

Identification

The colourful male is larger than the female. The laterally compressed body is ovate but long. The dorsal and anal fins are very long with the posterior end drawn out to a point in males. A number of oblique iridescent bluish stripes are present on the flanks, while the female is grayish and has larger bands along the body. Green and orange varieties are also available.

Breeding

Breeding methods are like all Gouramies. Male builds bubble nest. The bubbles are coated with mucous secretion to make them firm. The gravid female releases eggs under the nest and the eggs float upward towards the bubble nest.

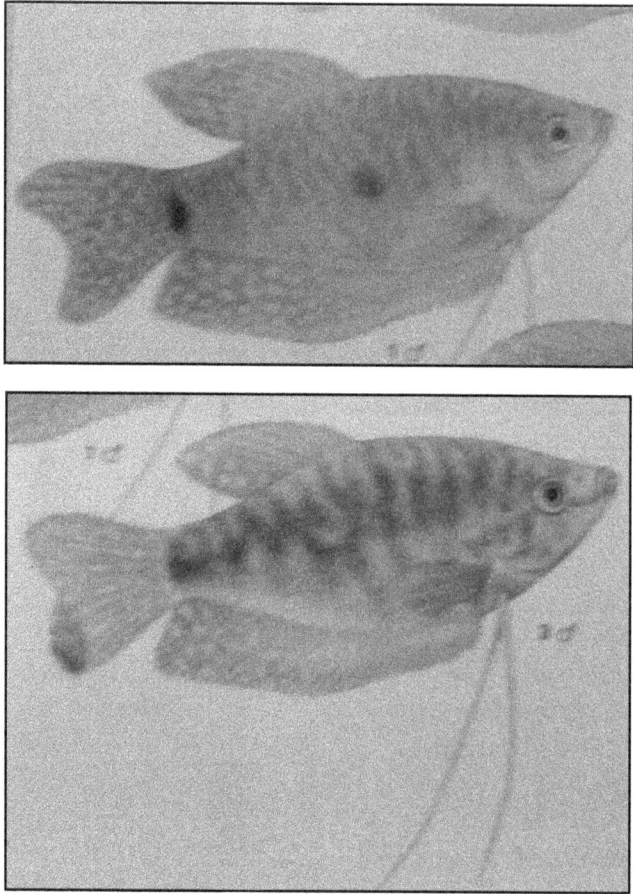

Figure 7.G.8: Indian Gourami

Thick Lipped Gourami

Common Name : Thick lipped Gourami

Scientific Name : *Colisa labiosa*

Identification

They exhibits sexual dimorphism. The male is longer than the female. They are colourful with the end of the dorsal fin pointed while female are less coloured with a rounded dorsal fin.

Breeding

Breeding method is like other Gourami.

(G) Hatchet Fish

Family	:	Characidae
Native	:	They are native of South America.
Species	:	There are two varieties of hatchet fish.

1. *Carnegiella strigata strigata*
2. *Gasterolepcus sternicla*

Carnegiella strigata strigata (Figure 7.G.9)

Common Name	:	Marbled hatchet fish.

They are having subspecies *Carnegiella strigata fasciata*

Identification

It is also known as marbled hatchet fish. It is one of the most popular among hatchet fish. They are deep bodied, silver purple in colour with greenish tint and dark broken lines that cross the lower body. The distinguishing characteristic of hatchet fish is the triangular body and mouth that is placed at the top of its head.

Water Quality

Temperature	:	24–28° C
pH	:	5.5 to 6.5 (slightly acidic)
Hardness	:	They require medium to hard water (5° dGH)

Figure 7.G.9: *Carnegiella strigata strigata*

Aquarium Behavior

They enjoy moving in shoals, hence it is suggested to have at least ten or more as a small shoal in aquarium. The aquarium should be covered as they like to jump out of water. Floating plants should be provided for shade.

Behavior with Other Aquarium Fish

These fishes are very delicate and should not be kept with any aggressive fish. Since they are surface dwellers, it is advisable to have discuss or angel fish in a community tank for middle water layer and corydoras catfish for lower layer.

Sex

It is difficult to differentiate sex but if viewed from the above the water and the female is fatter.

Maximum Size

They attain a size of 4 cm.

Breeding

Breeding is simple add peat extract to darken the water until it is almost opaque and provide subdued lighting. Feed them with small flying insects to induce spawning.

After extended courtship the female will deposit her eggs on plants. The eggs will hatch after a day and within 5 days the fry become free swimming.

Feed for Fry

The fry must be fed finely powdered flake food for the first three to four days, followed by baby brine shrimp.

Feed for Adults

They eat all kinds of live, flake or fresh food. However, floating flake or floating food is recommended to keep them in good nutritional balance. Depending up on the size of fish, Cyclops, Daphnia, blood worms and small brine shrimp can be given.

Gasterolepcus sternicla (Figure 7.G.10)

It is also known as silver hatchet fish. They are the native of Brazil, and Guyana. They are found in smaller streams with plenty of vegetation.

Identification

The distinguishing characteristic of hatchet fish is its triangular body with an upward pointing mouth, straight dorsal line and a dorsal fin positioned near to the caudal fin, the fish are surface oriented. White and black spots are found all over the body. White line can be seen near opercula.

Water Quality

Temperature	:	23–28°C
pH	:	6 to 7 (slightly acidic)
Hardness	:	They require medium hard water (2 to 15° dGH).

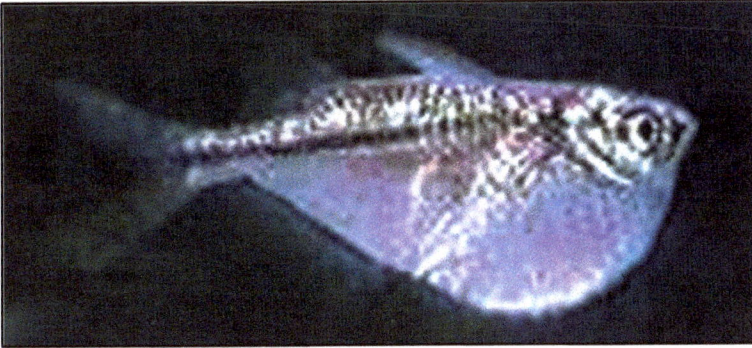

Figure 7.G.10: *Gasterolepecus sternicla*

Aquarium Behaviour

They are generally regarded as good community fish. They enjoy moving in shoals, hence it is suggested to have at least ten or more as a small shoal in aquarium. The aquarium should be covered as they like to jump out of water. Floating plants should be provided for shade.

Behaviour with Other Aquarium Fish

These fishes are very delicate and should not be kept with any aggressive fish. Since they are surface dwellers, it is advisable to have discuss or angel fish in a community tank for middle water layer and corydoras catfish for lower layer.

Sex

Males are slimmer than females.

Maximum Size

Their maximum attainable size is 6 cm.

Breeding

This fish is not successfully bred in captivity, unlike marbled hatchet fish.

Feeding

They will feed on mosquito larvae.

Tetra

Classification

Phylum	:	Chordata
Class	:	Osteichthyes
Order	:	Cypriniformes
Family	:	Characidae

Introduction

Members of this family amount to more than 1000 species. Most of the species come from central and South America. Most common feature of the characins are the presence of teeth and an adipose fin. It is always advisable to reproduce in aquarium the conditions similar to natural surroundings. They require soft acidic water of temperature about 20° C. The name "Tetra" evolved from the subfamily *tetragonopterianae*.

Commercially known Tetras are as under:

1. Cardinal Tetra (*Oaracheirodon axelrodi*)
2. Black widow Tetra (*Gymnocorymbus termetzi*)
3. Red eye Tetra (*Moenkhausia sanctaetilomenae*)
4. Neon Tetra (*Paracheirodon innesi*)
5. Black neon Tetra (*Hyphessobrycon herbertaxerodi*)
6. Heart Tetra (*H. erythrostigma*)
7. Lemon Tetra (*H. pulchripinnus*)
8. Serpae Tetra (*H. collistus*)
9. Blood fin Tetra (*Aphyocharax anisitsi*)
10. Glow light Tetra (*Vheirodon erythrozonus*)
11. Head and Tail light Tetra (*Hemigrammus ocellifer*)

Details of the species are as under:

Cardinal Tetra

Common Name	:	Cardinal Tetra
Scientific Name	:	*Cheirodon axelrodia* (Schulf 1956)
Native	:	Venezuela, Colombia and Brazil.

Identification

Cardinal tetras have uninterrupted horizontal streaks of electric blue and red on its flanks. It grows to 4 cm size.

Water Quality

They prefer acidic water (pH–5.5 to 6.00). Temperature around 26° C in breeding tank.

Food and Feeding

Live as well as artificial food.

Breeding

Female fishes can be identified by the presence of a fuller belly. Spawning is induced by a rise in atmospheric pressure. They spawn only when they kept together for 2 to 3 weeks. Spawning takes place at down or dusk. To induce spawning, addition of peat extract is advisable. The developing embryos are photophobic as in Neon Tetras. Eggs are usually laid in evening. Parents must be removed after spawning. Spanners are conditioned by giving live brine shrimp nauplii and kept in dimly light tanks with floating plants. The eggs hatch in a day's time and the newly hatched fry remain behind plants thickets. The fry can be nourished by providing Artimia nauplii and the later on with mosquito or chironomid larvae as they reach 1 cm in size. Cardinal tetras are very active and move in schools.

Black Widow Tetra

Common Name	:	Black Widow Tetra
Scientific Name	:	*Gymnocorymbus termetzi* (Boulenger, 1895)

Identification

When young the fish is very attractive with the posterior half of the body and large anal and caudal fins black in colour. The body also possesses two distinct vertical incomplete black bars across the abdomen. This fish grows to 7 cm in wild, but in the aquarium it grows to just 4 to 5 cm. The males are darker in colour with elongated dorsal fins. The female is a little wider and its anal fin is narrower and runs parallel with its second vertical line

Water Quality

They prefer soft water of pH 6 to 8.3 with temperature around 24° C.

Food and Feeding

Their main food comprises of algae and other live organisms, but they readily adopt themselves to prepared food.

Breeding

Several varieties are being bred but the most common has a long fin. This fish can easily breed in the aquarium. It is advisable to use spawning grids, but in its absence a well planted tank. They prefer diffused light. They are peaceful and could be kept in community tank. Spawners need to be conditioned by giving live food, particularly brine shrimp nauplii or frozen blood worms.

Serpae Tetra

Common Name	:	Serpae Tetra
Scientific Name	:	*Hyphessobrycon serpae* (Durbin 1908)

Identification

It has a glowing red body with yellowish abdomen. When in good condition, the posterior part of the body is blood red in colour. An elongated black blotch is present above the belly. The fish grows to 4.5 cm in size. Males are bright red in colour with belly less curved than in female. They are very peaceful fish which live in schools.

Water Quality

For breeding it needs slightly acidic water of temperature around 25° C.

Food and Feeding

They are omnivorous feeder and can easily adjust to the artificial feeds.

Breeding and Hatching

They are egg scatters, the minute gray coloured eggs hatch in 24 hours. The fry can be fed with Artemia nauplii.

Head and Tail Light Tetra

Common Name	:	Head and Tail Light Tetra
Family	:	Characidae
Scientific Name	:	*Hemigrammus ocellifer*
Native	:	French, Guyana

Aquarium Behaviour

They are peaceful and ideal for keeping in community tank. They prefer well–planted tanks with dark bottom and move in large schools.

Water Quality

Temperature	:	24–26° C
pH	:	6.3 to 7.0

Breeding

It is easy to breed. Breeding method is like other tetra.

Red Eye Tetra or Glass Tetra

Common Name	:	Red Eye Tetra or Glass Tetra
Scientific Name	:	*Moenkhausia sanctaefilomenae* (Steindachner 1907)

Identification

Red eye Tetra is a famous for the flashy red colour in the upper half of the iris. The body is gray with the edge of scales black in colour, giving the fish a local appearance.

Full Length

They grows to 7 cm in size.

Water Quality

Red eye Tetra requires soft acidic water with temperature around 24° C.

Feed

The fish is omnivorous in nature, consuming both live animal and plant material. Once is while provided them with Deck weed (Lemna).

Breeding

Breeding these fishes in aquarium is difficult. Females have more rounded and curved belly. The breeding pair has to be conditioned two weeks prior to spawning.

The male chases the female to plant thickets where she releases the eggs. Eggs hatch in 24 hours.

Feeding to Fry

The fry food feed on infusoria in early stages, followed by *Artemia* nauplii and Cyclops.

The young ones grow very slowly. The aquarium should be dimly lit with one-third vegetation and rest free-swimming space; it can be kept with other schooling fishes due to its peaceful nature.

Neon Tetra (Figure 7.G.11)

Common Name : Neon Tetra

Scientific Name : *Paracheirodon innesi* (Myers 1936)

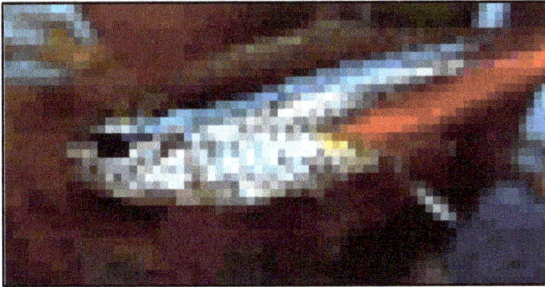

Figure 7.G.11: Neon Tetra

Identification

The body of Neon Tetra is coloured in shades of electric blue and red. The steak of electric red is incomplete at the abdominal region which is almost white in colour.. Females are plumper and a bend is visible in the electric blue line along the flanks. The colour part of their body is dark maroon and then has a thin bluish green stripe, extending the length of their body. Female is larger than the male and wider in proportion. The stripe is straight in the male while it is slightly crooked in female.

Full Length

They grow to 4 cm in size.

Water Quality

They prefer soft water of pH 6 to 6.5 and temperature around 21°C.

Breeding

Aquarium breeding is possible, but not easy. Prior to spawning, males and females have to be kept in separate tanks for two weeks in cooler waters of temperature 19° C. Only one breeding pair should be used at a time. Sterilized peat extract should be added to the breeding tank. Parents are to be removed as soon as spawning is over. Since the developing embryos are photophobic, the breeding tank should be properly covered and very dimly lit for the successful development of the eggs. Since young ones are very sensitive to light, the tank should be kept away from direct sun light until their eyes are completely developed.

Aquarium breeding is possible in well-planted tanks with diffuse light. Nearly 1000 eggs are released in one season. Periodic removal of water from breeding tank facilitates easy growth of fry.

Feeding to Fry

The fry starts feeding in 4 days and should be supplied with *Artemia* nauplii.

Aquarium Preference

They prefer heavily planted aquaria with reduced light to breed.

Hockey Tetra or Hockey-stick tesas

Common Name : Hockey Tetra or Hockey- stick tesas

Scientific Name : *Thayoria boehlkei* weitzman 1967

Identification

Light pinkish coloured fish is having black hockey-stick type thick line starting from gill cover to the end of lower part of caudal fin. The fish swim in a slanted position, titled to 30° upwards from the horizontal level. The hockey stick Tesas are native of Brazil and Peru. The name is based to a black like which runs through the entire body and continues down the lower lobe of the caudal fin. In a closely related fish, *Thayeria oblique* (Penguin fish), this black line extends only up to the second dorsal adipose fin. They are omnivorous fish. Females are with fuller belly.

Full Length

6 cm

Food

Live as well as artificial foods.

Sexual Dimorphism

The male smaller, slimmer, the female fuller in the body. Apart from the spawning period there is practically no difference between the sexes.

Sex Ratio

1 male : 3 female.

Breeding

Female lays more than 1000 small and brown coloured eggs during spawning period. After spawning the water should be replaced by freshwater of the same composition and temperature.

Incubation Period

12 hours.

Feeding of Fry

Up to four days fry use endogenous nutrition. The first five days *Paramecium caudatum* should be given one feeding at night in dim light. Then brine shrimp nauplii. Young fish grow rapidly.

Black Neon Tetra

Common Name	:	Black Neon Tetra
Family	:	Characidae
Scientific Name	:	*Hyphessobrycon herbertaxelrodi*
Native	:	South Africa

Identification

They are peaceful and pose no threat to any other fish in aquarium and are ideal for community aquarium. Although called the Black Neon, it is not a neon Tetra, but a completely different Species. General length of this fish is about 4 cm.

Food

Live as well as artificial foods.

Water Quality

Temperature	:	24–26° C
pH	:	5.5 to 6.8

Breeding

Breeding is to be done in breeding tank of 10 liter capacity with spawning grid. Lighting markedly dim, finely leaved plant should be provided. Breeding is reported to be difficult as live feed and soft acidic water is required to stimulate the fish to breed.

Sex ratio	:	1 male : 1 female
Eggs	:	They are small and glossy.
Incubation period	:	18 hours.

Feed of Adult

It feeds on flake food, adult brine shrimp, blood worms and other live food.

Feeding to Fry

First three days on *Infusoria (Paramecium caudatum)*. Monoculture or the finest freshly hatched brine shrimp nauplii.

Sexual Maturity Age

The fish reach sexual maturity at the age of eight months. Before spawning, the males and females should be separated for 14 days. The temperature of water should be kept at 20 to 22° C. Ripe females are not strikingly pumped. Breeding pairs are very shy and reluctant to spawning the new environment; spawning takes place after seven days. After the fifth day of hatching, the fry switch to normal nutrition. Their growth is rapid.

Aquarium Behaviour

They are peaceful and pose no threat to any other fish in the aquarium and are ideal for a community aquarium fish of similar disposition.

Bleeding Tetra

Common Name	:	Bleeding Tetra
Family	:	Characidae
Scientific Name	:	*Hypessobrycon erythrostigma*
Native	:	Amazon River

Identification

The male has an extended black dorsal fin.

Water Quality

Temperature	:	24 to 28°C
pH	:	5.7 to 7.2

Breeding

Breeding of this tetra in captivity has not been successful so far.

Feed

They feed on flake foods, adult brine blood worms and other live food.

Blood Fin Tetra

Common Name	:	Blood fin Tetra
Family	:	Characidae
Scientific Name	:	*Aphyocharax anisitsi*
Native	:	Argentina

Identification

The male has a small hook on its anal fin. The adult size is very small compared to other tetras.

Water Quality

Temperature	:	24–28° C
pH	:	6 to 8

Breeding

Spawners should be conditioned with live feeds like brine shrimp or blood worms. The parents should be removed after the eggs have been laid.

Glow Light Tetra

Common Name	:	Glow light Tetra
Family	:	Characidae
Scientific Name	:	*Cheirodon erythrozonus*
Native	:	Essoquibo River in Guyana

Breeding

It is reported to be difficult to breed. They spawn in warmer water (31°C) with medium hardness and pH 6.5. Only one breeding pair is kept at a time. Spawning takes place over several hours as one egg is laid at a time. Parents should be removed after spawning.

Breeding Preference

It prefers heavily planted poorly lit tanks. Water should be exchanged frequently.

Red Headed Tetra

Common Name	:	Red Headed Tetra
Family	:	Characidae
Scientific Name	:	*Petitella georgiae* Gery Boutiere 1904
Native	:	Peru, Hungallaga River, Upper Amazon River.

Identification

The red colouring is on the head (Some times on the flanks), the markings at the fin itself and on the anal fin. *P. georgiae* is more robust and the red colour on the head extends only to the gill cover and. There is a black mark above at the base of the caudal fin-one above and one below.

Full Length

Male–5 cm and Female–6 cm.

Food

Live as well as artificial foods.

Sexual Dimorphism

The male is smaller, more slender. They are egg layers.

Sex Ratio

1 : 1 set up several pairs.

Water Quality

Temperature	:	26° C
pH	:	6.5
Incubation period	:	24 Hours.

Feeding to Fry

First eight days on *infusoria* (*Paramecium caudatum*) monoculture, later the finest nauplii of brine shrimp or *Cyclops* nauplii.

Aquarium Behaviours

They can be kept in community tank.

Lemon Tetra

Common Name	:	Lemon Tetra
Family	:	Characidae
Scientific Name	:	*Hyphessabrycon Pulchripinnis*
Native	:	Shallow river of Brazil

Aquarium Behaviour

It is difficult to breed in captivity. It prefers warmer water 30°C and pH 6.5. Weekly change of water of about 25 per cent is recommended.

(H) Cichlids

Common Name	:	Cichlids

Classification

Phylum	:	Chordata
Super Class	:	Pisces
Class	:	Osteichthyes
Sub Class	:	Actinopterygii
Order	:	Perciformes
Sub Order	:	Percoidel
Family	:	Cichlidae

Due to their speciation, unique feeding habits and diverse characteristic in mating, the Cichlids have been a subject of research for last few decades. They are native of Africa and South America. The colouration in Cichlids is extremely variable and, as a rule, depends on the environment and the state of health of the fish. Young ones are somewhat more brightly coloured. In some species, the pelvic, anal and caudal fins are brightly coloured. In juveniles it is impossible to distinguish the sexes.

Species

1. *Pterophyllum scalar*
2. *Melanochromis auratus*
3. *Pseudotropheus lembarddoi*

Figure 7.H.1.1: *Pseudotropheus lombardoi* **(Male)**

Figure 7.H.1.2: *Pseudotropheus lombardoi* **(Female)**

General Information of the Species

Pair Making

Spawning is not difficult. They pair up normally on their own.

Spawning Method

Different spawning methods are observed in cichlids.

1. *Nest*: Male usually builds nest and also protect the eggs while female partner carries out all other defenses.

2. *Mouth brooding*: In mouth brooding cichlids, the female takes the fertilized eggs into her mouth from the bottom of aquarium. She also attempts to take in the "egg spots" or dummier, which the male has on this anal fin. By doing so, she sucks the sperm released by the male, which ensures fertilization of the eggs in her mouth. The female inside her mouth keeps the developing eggs until they hatch. She then releases the fry once they are free swimming stage. They are left to go free after fortnight.

3. *By cleaning spawning site*: Some species clean a suitable spawning site and lay their eggs on it. They protect their eggs.

Parental Care

Cichlids exhibit bi-parental care, with both male and female being involved. Both parents display territoriality and engage in mutual co-operative care of the young ones.

In mouth brooding, when fry are free swimming, she releases fry, but continues to protect the young ones when she feels insecure or senses danger or intrusion, by swallowing them in to her mouth for safe keeping.

When the eggs are on the substratum, they show degree of parental protection.

Food

Cichlids normally take live, dry frozen or other foods that can be offered in sufficient quantity. However, it is suggested to have flake food to start with and later feed with pellets supplemental occasionally with shredded shellfish, chopped earthworms etc. as they grow.

Details of Some Species of Cichlid

Nyasa Golden Cichlid

Common Name	:	Nyasa Golden Cichlid
Scientific Name	:	*Melanochrimis auratus*

Identification

The colouration in cichlids is extremely variable and as a rule, depends on environment and state of health of the fish. The male is velvety black with white longitudinal bands. The female is yellow with dark bands. Dark longitudinal black band is also found on dorsal fin.

Figure 7.H.2: *Melanochrimis auratus* **(M and F)**

Full Length	:	Male: 11 cm; Female: 9 cm
Diet	:	Live as well as artificial foods augmented by plant foods.
Sex ratio	:	1 male : 3–4 females.

Breeding

Eggs are incubated in mouth of the female lasts 22 to 24 days. The fry leave the mother's mouth when they are 1 cm long at which time they are fully self-sufficient and fully coloured. The litter usually consists of about 40 young ones. Breeding water should be fresh and filtered with following quality.

Water Quality

Temperature	:	26° C
pH	:	7.5 to 8.0
Hardness dCH	:	2–4 °

They should be kept in mono species tank spawned out females may be identified by the enlarged throat sac on the underside of the mouth. Transfer each female carefully to separate tank of about 20 liter capacity. The individual tank must be placed in a quite, undisturbed spot and provided with diffused light and aeration. The fry no longer stay by the female for any length of time and instinctively look about for a hiding place. The female may therefore, be removed. First of all grasp the female lightly in the palm of the hand, turn her head down ward and with a blunt rod lightly press on the lower jaw until mouth is opened. If there are still any remaining fry inside the female will spit them out.

Blue Acarea

| *Common Name* | : | Blue Acarea |
| *Scientific Name* | : | *Aegruidens latiforms* |

This fish is also known as blue spot cichlid. They are native of Panama and Columbia. It is one of the best-known cichlids.

Colour

The body colour is mainly a greenish blue with darker bands vertically positioned along the sides.

Adult Size

Fully grown specimen attain a length of 6 inches.

Maturity

They are mature when only about 3 inches long.

Angelfish

| *Common Name* | : | Angelfish |
| *Scientific Name* | : | *Pterophyllum scalar* (Lichtenstein 1923) |

Identification

Angelfish body is laterally compressed. It has a flat, upright disc like body with long dorsal and anal fins, long pectoral fins and widely splayed tail. Usually three numbers of vertical bands found on the body and fins with thinner and paler bars in between them. The fish has two feelers in front of the anal fin. The tail is vertically oriented and may be from the scoop shovel shape to long and relatively narrow depending variety. The fully-grown healthy specimen has a majestic and appealing appearance which justifies its name as angelfish *i.e.* angel among the ornamental fishes.

Commercial Available Varieties

The commercial available varieties of angelfish are as under:

1. Silver angel
2. Silver veil tail angel
3. Black angel
4. Black veil tail angel
5. Black lace angel
6. Ghost angel
7. Diamond angel
8. White angel
9. Marble angel (Two type)
10. Blushing angel
11. Zebra angel
12. Zebra lace angel
13. Golden marble angel
14. Koi angel
15. Smoky angel
16. Half black angel
17. Blusher
18. Sunset blusher
19. Germen blue blusher
20. Pandas
21. Albino angel
22. Pearl angel (Two type)

and many other recorded varieties.

Size

Generally it can grow up to 5–6 inches long. The top and bottom fins spanning a greater distance in veil tail varieties.

Life span

Angelfish can live over 10 years with the provision of nutritionally balanced feed, ideal optimum water quality and maintenance of better sanitary and hygiene. But generally it can survive 5–6 years.

Characteristic of Angelfish

1. Compatible fish and very much suitable for community tank.
2. Very hardy and sturdy fish and easier to acclimatize in the water environment.
3. Swim all levels of aquarium and therefore, occupies any level of the tank.
4. Omnivorous fish, which can take wide variety of food.

Different Varities of Angel

Figure 7.H.3.1: White Angel

Figure 7.H.3.2: Blushing Angel

Figure 7.H.3.3: Kole Angel

Figure 7.H.3.4: Zebra Lace Angel

Figure 7.H.3.5: Silver Angel

Figure 7.H.3.6: Koi Angel

Figure 7.H.3.7: Golden Marbel Angel

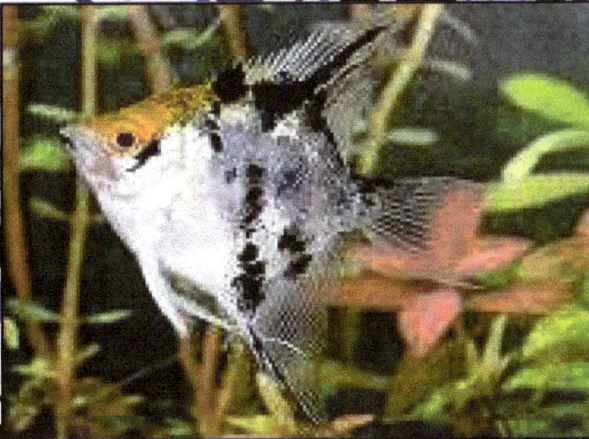

Figure 7.H.3.8: Yellow Head Angel

5. Becomes tame very quickly with the aquarium fish keeper.

6. Available across the global particularly all tropical countries.

Identification of Sex

Selection of breeding pair is first step in the spawning of angelfish. Knowing the sex is very difficult and it is virtually impossible for the beginners to identify the male

Figure 7.H.3.9: Zebra Angel

Figure 7.H.3.10: White Marble Angelfish

Figure 7.H.3.11: Black Angelfish

Figure 7.H.3.12: Angel

Figure 7.H.3.13: Angel Fish

Figure 7.H.3.14: Golden Angel fish

and female in the young stage. During breeding season some of the characters found in the body features as described below:

Character	Male	Female
Abdomen	Flat	Swollen
Papilla	Pointed	Rounded
Vent	Small	Larger

However, even by these characters much experience is required for identifying male and female.

Selection of Breeding Pair

The most logical approach for selecting at least six or more healthy and active fishes of about 4–5 months age or less can be purchased from reputed dealers. Under favourable conditions the angelfish reach the breeding size within a year or so and at least one pair is hoped form the lot of six or more fishes.

When two angelfish start pairing it is generally seen that they protect a certain area in the aquarium. Young male as well as female often spread their fins and gills dancing in a half circle in form of their mate. Another sign of commencement of breeding and pairing is that the male angelfish will dart at fish other than its mate in order to establish a territory.

After getting the pair the male and female brooders are to be immediately removed from the community tank and then kept in a separate tank of approximately 100 liter size. They are conditioned in the breeding tank are supplemented with high protein balanced diet optimum water quality and better hygiene is to be maintained in the tank.

Water quality for successful breeding and for better growth and survival.

Temperature	:	24–26° C
pH	:	6.5–6.9
Dissolved Oxygen	:	5–6 ppm
Hardness (mg CaCO$_3$/Lit)	:	75–100 ppm
Free Ammonia (NH$_3$)	:	Less than 0.1 ppm
Nitrate Nitrogen (NO$_3$)	:	Less than 0.01 ppm

Spawning

The female prefers to lay eggs on the spawning site. If spawning site are not available then it can develop a habit of laying eggs on the walls and aeration tubing. This is to be discouraged by providing plastic plants, used pots, polypipes or plain slate rectangles.

The most convenient substrate is slate or acrylic plate of size 24 × 10 inches with base and support. This should be kept in an inclination position in the wall of the tank.

Eggs are generally sticky in nature. The pair first selects the spawning site and thoroughly cleaned the site before the spawning takes place. The large healthy female lays up to 1000 eggs in 2 hours but the general average fecundity is 400–500 eggs. The female lays her eggs in a line on the substratum and then male deposits its sperm over it and fertilize the eggs.

Breeding Stimulation

The breeding stimulation is enhanced by raising the water slowly and also doing partial exchange in breeding tank (about 25 per cent of total water volume). Unfertilized eggs become white within 24–48 hours of egg laying.

Parental Care

The parents particularly female continuously fan the eggs by its pectoral fin. The fanning serves two purposes *viz.* firstly, it clean the eggs and does not allow external pathogens like fungi or protozoan to grow over the eggs and secondly, it supply sufficient oxygen to the eggs by means of fanning.

Hatching

It is better to remove the fertilized eggs and kept in a separate hatching tank for better hatching. The eggs generally hatched after 72 hours. Sometimes the eggs and freshly hatched young ones are susceptible to external pathogens.

If they are attached by pathogens the safer fungicide like acriflavin, malachite green, copper sulphate, potassium permanganate and formalin are to be used with proper dosage.

After hatching the fry are sticked together to the substratum for about 1–2 days before becoming free swimming. They attained the free swimming stages after 3–5 days.

Food for the Fry

The fry do not require any feed up to the free swimming stage and once it attained the free swimming stage the fry should be supplemented with live feed like brine shrimp. Daphnia, moina, blood worm, tubifex worm, rotifer and infusoria etc. for at leat 2 weeks. The required feed can be given in split dosages 5–6 times daily.

They may be fed with commercially available freeze-dried food such as spirulina, blood worm, tubifex worm in the form of powder initially and granules and pelleted feeds subsequently as they grow in the bigger size. The fry rearing tank should be thoroughly cleaned twice daily in order to remove the unconsumed excess feed and excreta of fishes. During siphoning the excess feed, care must be taken not to remove or vacuum up the fry.

Growth

Generally it takes 2.5 to 3 months for the fry to become the size of a quarter. Although the life span of fish recorded up to 10 years, the fish fertility of the male and female remains up to 3–4 years. After that the quality of eggs and milt are not good enough to produce the healthy young ones.

Flower Horn

Common Name	:	Flower Horn
Family	:	Cichlidae
Scientific Name	:	It is hybrid species of *Cichlasoma cichlids*
Native	:	Asia, Eastern Africa and South and Central America

Identification

Though this is hybrid species, it has many differences from "Cichlids" in its colour, appearance and even hump shaped forehead, it retains several similar characteristics such as aggressive and territorial behaviours.

They are more than a dozen varieties of Flower Horns in market.

1) Oriental Beauty

It has the most attractive markings on the body. These markings connect closely from the tail to the gill and sometimes extend to the head. It has a sickle shaped back fin and a semicircular tail. As the shining scales resemble twinkling stars, this variety fetches a higher price and is believed to bring luck and prosperity. The Flower Horn is considered to be a good "Feng-Shui" fish. A single fish of this variety can fetch a price up to US$ 3000.

2) Galaxy Blue

This fish shows its conspicuous colours in the dark. Its beautiful colours make it look like a hunter on mysterious night. It has irregular strips on the body and a pair of red eyes distinct from other varieties. Normally it has perfect hump head a cherry–like mouth and a neat straight back fin.

3) Wonder Spark

With blue shining dots this varieties looks like an ordinary Cichlids. It is also known as the "Panther Warrior" as it is the fastest swimmer among the Flower Horns. It has unique sequence of horizontal marking and lightening dots all over the body. (Figure 7.H.6)

4) Strom Rider

This variety has hard and solid scales, making it appear to be wearing a coat of mail. There are seven different sizes of marking on the side and five vertical strips. (Figure 7.H.5)

5) Happy Star

This variety has a most colourful pattern on the whole body it had red shining eyes and golden face. The abdomen is blue and red and body has eight clusters of markings. (Figure 7.H.4)

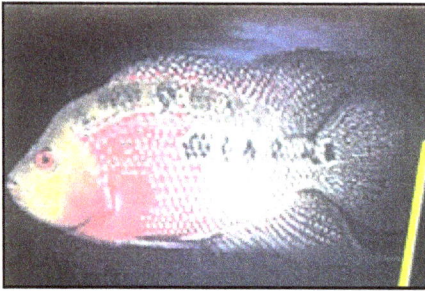

Figure 7.H.4: *Cichlasoma* spp. "Happy star"

Figure 7.H.5: *Cichlasoma* spp. "Strom Rider"

6) Royal Tiger

It is so named because of the strips on its body. It has a dark red and pink body, golden face and blue dotted scales.

7) Red Beauty

This variety has a most striking red colouration. It has two pairs of marking on the side of the body and blue dots in the rear and extending to the tails.

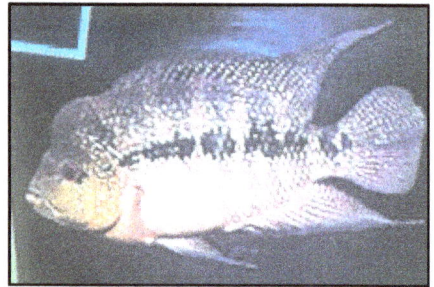

Figure 7.H.6: *Cichlasoma* spp. "Wonder Spark"

8) Moon Light Beauty

As the name indicates, Moon light Beauty belongs to the beauty of night with seven vertical indistinct bands and horizontal clusters of markings on the body. Moon light Beauty exhibits varied colouration during the night.

9) The May Blossom

This is very ordinary unlike other varieties of Flower Horns. Due to light colouration, it is also known as "Light Blossom". It has light red and yellow body colour with six clusters of distinct horizontal markings.

Sex

They are egg layer fish. The male can grow up to 12 inches or more. Matured males have a characteristic hump head and are larger than females. The forehead of the female is less conspicuous and tapers towards the mouth.

Water Quality for Breeding

Flower Horns need space and clean water to stay in good health. The water hardness is required. pH should be between 7 and 8. The tank bottom should have gravel with which they can build their nests.

Breeding

Flower Horns are easy to bread, adapt to new environment quickly and establish a friendly relationship with their care taker. They are easy to pair. During courtship, the pair will make a small depression on the gravel bed and use it as a nest.

Fertilization

Fertilization is external and more than 1000 eggs are laid at a time. Survival rate can be as high as 70 per cent. The eggs may remain attached to the gravel until they are independent or swimming, which under normal condition will take about a week.

Feeding to Fry

During the initial stages live artemia nauplii is ideal as food.

Julidochromis ornatus

Family	:	Cichlidae
Scientific Name	:	*Julidochromis ornatus*

Identification

They are known as cave spawners. Although these cichlids are available in market they are not well known among many of the hobbyists due to their higher value and the fact that they are not as productive as other cichlids and growth to reach marketable size takes several months.

Water Quality

Temperature	:	22 to 28° C with an ideal average of 25° C
pH	:	8–9
Hardness	:	8–15 dH and should not drop below 8 dH.

Breeding

They are coming under group cave Spawners or shelter breeders. Although these fishes breed in community tanks, success is often seen when they are kept as a single breeding pair. No particular water condition is required for spawning, provided water quality is good and pH is maintained within the desired range. Females guarding egg or fry may become aggressive.

Aquarium Behaviour

Plants are not recommended in aquarium condition. It would be better to provide rock caves instead. They are territorial and peaceful.

Tropheus duboisi

Family	:	Cichlidae
Scientific Name	:	*Tropheus duboisi*

Tropheus duboisi are mouth breeders. They are originated from lake Tanganyika. Among Tanganyika cichlids, this species is popular among hobbyists. On olive green body white colour spots from dorsal fin side to stomach side are seen in horizontal line. White spots are also seen on dorsal fin.

Maximum Size

Maximum they can reach to 8 cm.

Water Quality

Temperature	:	22–28° C with an ideal average of 25° C
pH	:	8–9
Hardness	:	Should not be less than 8 dH, they prefer between 8–15 dH

They do not prefer sudden fluctuation in water quality or temperature. Moderate high carbonate hardness is recommended to keep pH stable.

Breeding

They are mouth brooders. Although these fishes breed in community tanks, success is often seen when they are kept as a single breeding pair.

Aquarium Behaviour

They are aggressive towards each other but considered less aggressive towards other fish. Aggression is rare and usually only occurs during spawning. They prefer rock caves in aquarium.

Discus

Common Name	:	Discus

Classification

Phylum	:	Chordata
Super Class	:	Pisces
Class	:	Osteichthyes

Sub Class	:	Actinopterygii
Order	:	Perciformes
Family	:	Cichlidae

Identification of Species

There are two species of Discus.

1. *Symphysodon acquifasciata*
2. *S. discus.*

Three wide prominent vertical dark bands characterize the later one on the center of its body, one across the face (through the eye) and one across the caudal peduncle. The bands between these dark bands are usually much thinner and very faint.

S. acquifasciata, on the other hand, has eight uniformly wide dark vertical bars. If is subdivided in to three sub-species.

1. *Symohysodon acquifasciata* (The Green discus)
2. *S. acquifasciata hedri* (The Blue discus)
3. *S. acquifasciata axelrodi* (The Brown discus)

Water Parameter for Growing Discus

| *Temperature* | : | 26 to 30° C |
| *pH* | : | 6.0 to 6.4 |

Life Span

Under the optimum condition Discus will live for 10 to 12 years.

Ideal Composition in Aquarium

With a pair of *Corydoras* Cat fish and cardinal tetra Discuss can be kept in aquarium.

Selection of Breeding Stock

Maturity

The female Discus will attain its sexual maturity at the age of 10–12 months and the male Discus at the age of 14–15 months.

When they are sexually mature the female usually select her mate. The potential breeding pair will separate themselves from the other fish and settle down in a corner of the aquarium. They will then always stay together and defend their territory.

Feed for Breeder

The breeder fish should be fed regularly three times a day with live tubifex worms, blood worms and beef heart mixed with multi-vitamins.

Spawning

At the time of spawning, female leads the male to the tank; she will keep away from them. Once a pair is formed, they are transferred into an other spawning tank. To minimize the rise of the eggs being eaten by the parents, it is best to select fish of at least one and half year old.

Spawning Requirement

A breeding tank measuring 60 cm × 60 cm × 38 cm is big enough for one breeding pair to build their nest in. A piece of PVC plastic pipe of 10 cm diameter and 20 cm height or an inverted flowerpot, can be placed in the tank as spawning substrate.

Water Quality

Following water quality may be maintained.

pH	:	6.0 to 6.5
Temperature	:	28–29°C
Water Hardness	:	46.2 mg/l CaCO$_3$

Once the pair has chosen a particular spawning site, the female fish carefully starts cleaning the site, occasionally, the female will swim towards the male, bow and return to continue the cleaning. The cleaning process is sure sign that spawning is imminent. The cleaning of the spawning site will normally take several hours or perhaps a whole day. As the activity around the spawning site becomes more vigorous, it should become possible to distinguish the sexes.

Sex

In the male, the breeding tube is smaller, shorter and pointed. While in female, the breeding tube is thicker, blunt and projected further from the body.

After a number of "trial runs", the female swims towards the spawning site and lay one egg after another. She will deposit on an average of 10–20 eggs in a single raw. After the female has laid the eggs, the male will swim over them, depositing his sperm over the eggs. The sequence of events will take about 2–3 hours. The number of eggs will depends on the age and condition of the female.

Fecundity

There may be as few as 30 or as many as 500, but the average seems to be around 200.

Hatching

Depending on the temperature and other water conditions the eggs take 5–6 hours to develop. Hatching during darkness would be fatal for all of the off spring, because the parents would not be able to gather them together. If the water conditions are unsatisfactory for the brood, the larvae will die. A full flagged little discus fry can be seen in the water usually five to six days after fertilization.

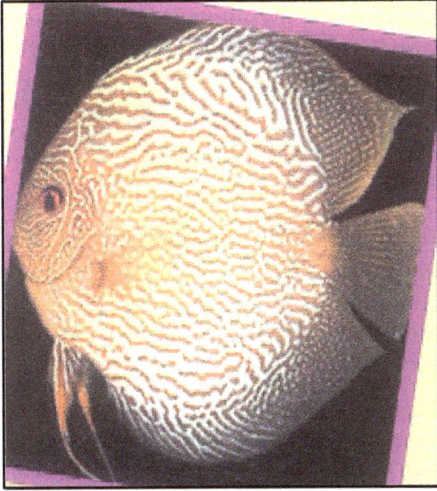

Figure 7.H.7: (Discuss) Pigeon Blood

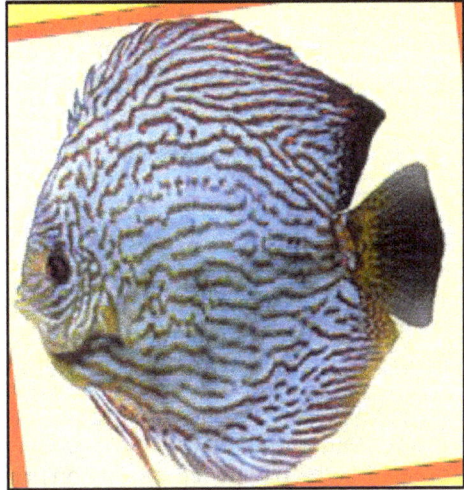

Figure 7.H.8: (Discuss) Blue Turquise

Figure 7.H.9: Red Turquoise

Figure 7.H.10: Brown discuss

Satisfactory Water Condition for Brood

pH	:	5.0 to 6.0
Water Temperature	:	28–29° C
Nitrate and Nitrite	:	Very law
Concentration of Germs	:	Very law.

Parental Care for the Fry

During the hatching period, the parents will be fanning the eggs as well as picking off the dirt and eating any eggs that do not develop properly. When the eggs have hatched, the parents suck the larvae out of their eggshells. The larvae can be seen wriggling in a great black mass attached to the spawning site. Some time a larvae wriggles itself free of the spawning site and is usually picked up by one of the parents before it reaches the bottom of the aquarium. Sometimes the larvae are moved to another site by the parent fish. This is protective measure against any predation. During this period, they obtain their nourishment from their yolk sac.

Feed for Larvae

Seventy two hours after (depending on the water temperature) spawning, the fry become free swimming. They will swim aimlessly about the tank. Within an hour, all the fry can be seen grazing near the sides of the parents. They are eating the slime secretion from the parent's body. The slime secretion is vital to the young fishes for at least 4–5 days and they will not take any other food.

Newly hatched brine shrimp can be introduced to the fry after 5–6 days. Brine shrimp should be given as often as possible and it is better to give small quantities at a time. Fry which have fed heavily on brine shrimp nauplie or formula feed tends to be lazy and will just lie on the bottom.

The fry can be fed with egg yolk paste smeared in a thin strip around the rim of the tank. After four days, supplementary feeding of freshly hatched brine shrimp should be started. The gentle flow of water through the feeding tank helps keep fry swimming. For the first six months, they can be fed 5 times daily with brine shrimp and supplementary feed. After six months feeding can be reduced twice a day. Apart from the formula feed for Discus, live feeds like blood worm, tubifex worms, earthworms, brine shrimp and even frozen dried worm meal can be used.

A good feeding programme is very important during the first year of Discus. The young Discus is left with the parents for approximately 2–3 weeks. Thereafter they can be moved to another tank. At this stage water exchange should be done at a rate of about 20 per cent daily. Tap water that has been aged overnight is most commonly used.

(I) Loaches or Botia

Classification

Phylum	:	Chordata
Super-class	:	Pisces
Class	:	Ostoichthyes
Sub-class	:	Actinopterygii
Order	:	Cypriniformes
Super order	:	Ostariophysi
Family	:	Cibitidae
Native	:	They are native of India, Thailand, Pakistan, China, Bangladesh, and some Indonesian island.

General

There are around 40 species of loaches known today belonging to the genus *Botia*. Loaches are predominantly bottom dwellers and are often found under stones or pieces of driftwood. They are probably the most diverse group of fishes in the hobby both in patter and behaviour ranging from extremely peaceful *Botia histronica* to the aggressive *B. beaufforti*.

Important Species

1. *Botia macracanthus* (Clown loach)
2. *B. sidthimunki* (dwarf loach)
3. *B. almorhae* (Yo Yo loach)
4. *B. histrionica* (Golden Zebra loach)
5. *B. beuforti*
6. *B. dario* (Queen loach)
7. *B. hymenophysa* (Tiger loach)

Identification

Identification of some of the important species are given below.

1. *Botia macrocanthus* (Clown Loach) (Figure 7.I.1)

It is available in Sumatra, Indonesian Borneo. It is most colourful and popular among the loaches. They are moderately elongated and, laterally compressed in shape with a small arched back, straightly belly profile and four pairs of small barbells leached on the lower jaw. They look beautiful in the aquarium. The head is large and the mouth faces downwards with fleshy lips. The spine in front of the eye is quite short. The body colouration is bright orange and three wide, wedge shaped black bands cross the flanks. The first band runs from the top of the skull across the eye and then obliquely down to the region of the mouth; the second starts in front of the dorsal fin and extends down to the belly, while the third covers a large part of the caudal peduncle and runs down to the anal fin. The pectoral, ventral and caudal fins are yellowish with black margins.

Food

They can be easily kept wide variations of food including brine shrimp, bloodworms and snails.

Figure 7.I.1: Clown Loach (*Botia macrocanthus*)

Water Quality

They prefer soft water, which needs to be changed often. They can be kept in community tank.

2. *Botia sidthimunki* (Dwaft Loach) (Figure 7.I.2)

They are native of Thailand; however, it is reported to be extinct in its native habitat due to over fishing and water pollution.

Identification

Juveniles carry a chain-link pattern on the body, which is why it is also known as the chain loach, while the adults or brooders develop long thick black strips along both sides.

Figure 7.I.2: Dwarf Loach (*Botia sidthimunki*)

Water Quality

pH	:	6.2 to 7 (Slightly acidic)
Temperature	:	24 to 27 °C

Breeding

Their spawning behaviour is known, breeding in captivity has not been achieved so far.

3. *Botia almorhae* (Yo Yo Loach)

It is also known as Pakistani loach. It is originates from India and Pakistan and is found in slow flowing waters.

Figure 7.I.3: Yo Yo Loach (*Botia almorhae*)

Behaviour

They will quickly become accustomed to captivity and remain in sight during daylight hours although most of the botia are considered to be nocturnal.

4. *Botia histrionica* (Golden Zebra Loach) (Figure 7.I.4)

They are native of India, China and Myanmar and also known as Burmese loach. They are very active among other botias and are generally very peaceful and social species.

Figure 7.I.4: Golden Zebra Loach (*Botia histrionica*)

Identification

They are characterised by a short nose, black bands on the body that disappear at the bottom of the body. The bands extend in to the caudal fin as distinctive vertical strips, 2 or 3 on both lobes. The characteristic thinner black line extends through the eye to the mouth, a feature which helps distinguish this species in its juvenile stage from the very similar looking *B. rostrata* and *B. almorhae.*

5. *Botia beauforti* (Figure 7.I.7)

This is territorial and is known for its aggressiveness in nipping the fins of other fish in community tank.

Identification

Its red dorsal fin and dark red tail with black dots over a light grey-green body make is very beautiful. Generally, they are kept in groups of 3 to 5 to reduce aggression towards other fish. They are mostly nocturnal and need hiding places in the aquarium.

6. *B. Dario* (Queen Loach) (Figure 7.I.5)

It is one of the most active loaches originated from India and Bangladesh, and commonly known as the queen loach. With yellow golden strips on a black background, they look attractive. In young ones the bands are set further apart and become more numerous in adults. They enjoy resting hidden between dense plant leaves. They prefer live feed, but will accept flake and sinking food.

7. *B. hymenophysa* (Tiger Loach) (Figure 7.I.6)

Scientific Name	:	*Botia hymenophysa*
Distribution	:	India, Malay Peninsula, Greater Sunda Islands.

Figure 7.I.5: Queen Loach (*B. Dario*)

Figure 7.I.6: Tiger Loach (*B. hymenophysa*)

Full Length	:	20 cms
Food	:	Primarily live food (worms) may be augmented by artificial foods.

Compared with the other members of the genus *Botia, B. hymenophysa* has more elongated body with broader tail end, like *B. berdmorei*. The fish should be kept in a large community solitary specimen and groups on only a few fish are shy and quarrel some. Nothing is known about the manner of reproduction and sexual dimorphism.

Figure 7.I.7: Polka Dotted Loach (*B. beauforti*)

(J) Barbs

Classification

Phylum	:	Chordata
Super class	:	Pisces
Class	:	Osteichthyes
Sub class	:	Actinopterygii
Order	:	Cypriniformes
Family	:	Cyprinidae

Barbs are the members of the Cyprinidae family, comprise more than 400 species in their entire range. They are available in Asia, Africa, Europe to Central China, Philippines and the West Indies. Barbs are not recorded from Australia and South America. Out of about 150 species of barbs are of ornamental value.

Most Common Barbs

1. Tiger barbs (*Barbus hexazona*)
2. Rosy barbs (*B. canchonius*)
3. Cherry barbs (*Puntius titteye*)

Wild Caught Varieties

1. Sophores barb (*Punctius sopores*)
2. Terry barb (*P. terio*)
3. Tic-Tac-Toe barbs (*P. ticto*)
4. Shalini barbs (*P. shalynius*)
5. Denison barb (*P. denisoni*)
6. Melon barb (*P. melanamphyx*)
7. Sahyadri barb (*P. sahyadri*)
8. Aruli barb (*Barbus arulius*)

Identification of Common Barb

Tiger Barb

Common Name	:	Tiger barb
Scientific Name	:	*Barbus hexazona* (Mc Clelland 1839)

They are native of Sumatra and Borneo in Indonesia. It has a laterally compressed body about 6 cm long. The basic colour of the body is silvery white with brownish upper part of a green shine. Four bluish vertical bands running across the body characterize the tiger barb. The dorsal fin has a black base with a reddish fringe while the pectoral fin is reddish and the snout has a reddish tinge. The red pigmentation is less prominent in the female. The wild form is known as a black variety. Today, several varieties of tiger barbs are cultured through cross breeding.

Figure 7.J.1: Tiger Barb (*Barbus tetrazona*)

Black pigmentation is completely reduced in the gold variety and the fish is orange-gold in colour with four white coloured bands. While the moss green variety has a deep green coloured body with four dark bands.

There is a mutant variety cold ghost, where in the tiger barbs has became "Transparent". So that vessels in the gills and internal organs are visible. Black, gold and moss green varieties of ghost are also now available in the market.

Rosy Barb (*B. canchonius*)

Common Name	:	Flying Barb
Family	:	*Cyprinidae*
Scientific Name	:	*Esomus danrica*
Synonyms	:	*Nuri danrica*

Rosy barb is originally from West Bengal, India. Male rosy barb is red and female is yellow in colour with black spots near the rear and on the dorsal fin. Rosy barbs are schooling fish. The red or rosy colour intensifies during pawning. In all species of barbs the body is covered with relatively large scales while the head is without scales.

Figure 7.J.2: Rosy Barb (*Barbus conchonium*)

Flying Barb (*Esomus danrica*) (Figure 7.J.3)

Distribution

India, Sri Lanka, Thailand, Shallow mater with large surface area also rise paddies.

Identification

They are surface school fish that are accomplished swimmers and jumpers. They have highly developed pectoral fins, which enable them to leap above the surface for food (insects). They have one pair of long barb. One dark black line from mouth to tail and one golden line above the black line starts from gill cover to tail is the characteristic identification mark.

Figure 7.J.3: Flying Barb (*Esomus danrica*)

Full Length

15 cms.

Food

Live as well as artificial foods.

Black Spot Barb

Common Name	:	Black spot Barb
Scientific Name	:	*Putius filamentosus*
Synonyms	:	*Leuciscus filamentosus, Barbus mahecola, B. filamentosus, Systomus assimilis*

Distribution

Southern and South West India, Sri Lanka mountain streams.

Identification

It can be identified with big black spot on the body above ventral fin, on the lateral line. Caudal fin is indented (Bifurcated) with orange colour. Colour of dorsal fin is orange. Colour of fish is golden yellow.

Figure 7.J.4: Black Spot Barb

Full Length

15 cm.

Sexual Dimorphism

The male has the dorsal fin rays extended, terminating in a pectinate elongation, and a spawning rash on the snout above the upper jaw.

Sex Ratio

1 male to 1 female (Artificial fertilization is also possible)

Water Quality for Breeding

Temperature	:	26° C
pH	:	7.0
dCH max.	:	1°

Sexual Maturity

The breeder fish attain sexual maturity at approximately 18 months.

Breeding

Spawning takes place among plants at the surface and is very vigorous. During the extremely brief act of mating the male is pressed against the female and lightly entwines her body with his tail fin. When the spawning is over the parent fish should be removed from the tank and half of the water should be replaced by freshwater of the same composition and temperature. Colour the water faintly with methylene blue.

Feeding to Fry

Nauplii of brine shrimp on cyclops. Because of the great number of young fish make sure they have sufficient food and freshwater. Later sort them according to size to promote regular and rapid growth. Young fish have a typical juvenile colouration that is quite different from that of the adult fish. This colouration disappears at about seven months.

Long Fin Barb (Figure 7.J.5)

Common Name	:	Long fin Barb
Family	:	*Cyprinidae*
Scientific Name	:	*Capaeta arulia*
Synonyms	:	*Barbus arulius, Puntius arulius*

Identification

This fish can be identified with two thick bands, starting from beginning and from end of dorsal fin, which disappear near stomach region. Two dark black spot one on the body under the dorsal fin and second near caudal fin. Caudal fin is of indented type. One barb on the both side on upper jaw is found. Colour of fish is light pink with blackish tinge on lateral line. Slight greenish colour between eye and dorsal fin can be seen.

Full Length

12 cm

Sexual Dimorphism

The male is slimmer and slightly smaller, with dorsal rays lengthened in fan-like fashion and with a spawning rash round the mouth.

Sex Ratio

1 male : 1 female (The fish may also be bred in a shoal)

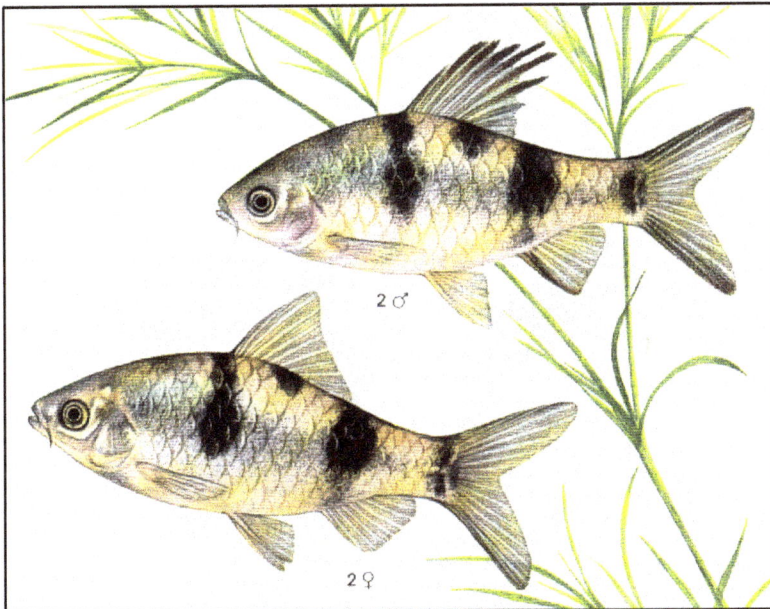

Figure 7.J.5: *Capoeata arullas*
(Male upper) (Female lower)

Other Species of Bark

Figure 7.J.6: Tic-TacToo Barb (*Puntius ticto*)

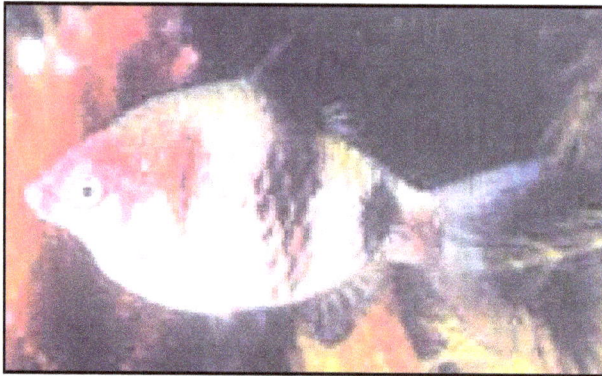

Figure 7.J.7: Black Ruby Barb (*Punctius titteye*)

Figure 7.J.8: Green Tiger barb

Figure 7.J.9: Tinfin Barb

Water Quality for Breeding

Temperature	:	24–26°C
pH	:	6.5 to 7.0
dCH	:	< 2°

Behaviour

They can be kept in a tank with 100-lit water per pair, spawning grid or finely leaved plants should be provided. Diffused light will help in spawning.

Eggs

Slightly adhesive, colour of eggs are like tea. Size are 1.5 mm. Incubation period 35 hours/26°C.

Feeding to Fry

Cyclops or brine shrimp nauplii. Seven days after becoming free swimming the fry attain a length of 3.5 mm. Their growth is relatively rapid.

(JD) Danio

Common Name	:	Danio
Family	:	Cyprinidae
Native	:	Indian peninsula, Sri Lanka, Pakistan, Thailand, Myanmar, Malaysia and Indonesia.

Habitat

They are found in a variety of habitats from boulder strewn mountain torrents to small pools in dry zone streams. They are mostly found in flowing water rather than reservoirs or tanks.

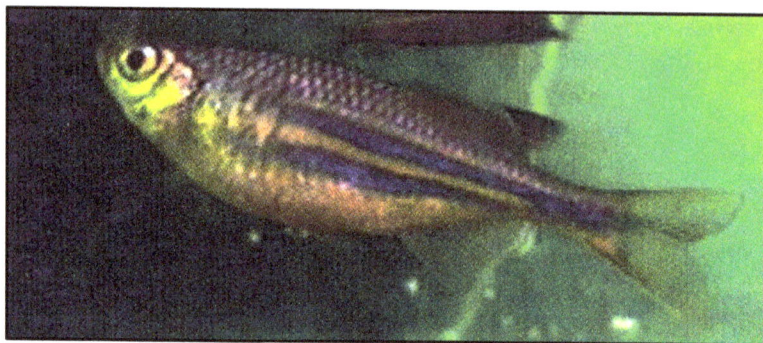

Figure 7.JD.1: *Danio malbaricus*

Species

There are more than 12 species reported today of which *Danio malabaricus, D. albolineata and Brachydanio rerio* are common in the hobbyist market.

Danio malbaricus

They are commonly known as giant danio. They are native of Southwest coast of India and Sri Lanka and have been widely transported around the world through the aquarium fish trade.

Identification

The head region is silvery coloured and rest of the body is olive green, decorated of yellow-cream to bluish reflections. The mouth is directed towards the surface. The narrow and elongated body, with colourless or transparent pectoral, dorsal and anal fins is diametrically opposite in the back half. The caudal fin has two lobs. The females have a rounded belly while the dorsal fin is longer; the pelvic fin is more pointed in males. The males are slimmer and have more colouring.

Full Length

The maximum size is 15 cm but generally they grow near to 12 cm.

Water Quality

Temperature	:	27° C
pH	:	Around 7

Food and Feeding

They are omnivorous and will accept any type of food, but prefer live or frozen brine shrimp, small insect and worms. They also feed detritus.

Sex

During reproduction, the male's pectoral fins become orange with reddish tinge. The blue central line is rectangular while in female this is curved upward at the base of the caudal fin.

Breeding

They spawn in shallow water among marginal weeds and roots, usually after heavy rain. They spawn easily and lay 200–300 eggs at a time which hatch in 1–2 days. The eggs are sticky and stick on plants. The young ones will start free swimming on the fifth day.

Care should be taken to remove the parents after spawning to prevent the eggs being eaten.

Aquarium Behaviour

They are very peaceful. They are gregarious in behaviour and prefer to live in groups of 5 to 7. They go well in community tanks and prefer flowing water

Pearl Danio

Common Name	:	Pearl Danio
Family	:	Cyprinidae
Scientific Name	:	*Danio albolineata*
Native	:	Streams and rivers in Thailand, Myanmar, Malaysia and Indonesia

Figure 7.JD.2: *Danio albolineata*

Danio albolineata

They are better known as Pearl danio. They are found in streams and rivers in Thailand, Myanmar, Malaysia and Indonesia.

Identification

This is the prettiest of all small danios and is found in the upper and middle layers. They are overall blue-silver in colour with a reflective pinkish red mid-lateral strip that runs from just behind the belly to the caudal peduncle. Males are more slender, small and brightly coloured than females.

Figure 7.JD.3: *Spotted danio*

Full Length

The adult size is 5 cm.

Water Quality

Temperature	:	25° C
pH	:	6.5 to 7.5 (Slightly acidic)

Food and Feeding

They are omnivorous and hence all types of foods are taken. Especially small insects are preferred.

Breeding

Breeding generally takes place under lower water condition. They lay sticky eggs on plants.

Aquarium Behaviour

They are active swimmers and often jump out from the aquarium when frightened. They go well in community tanks and prefer running water.

Zebra Fish or Zebra Danio

Common Name	:	Zebra fish or Zebra Dabnio
Family	:	Cyprinidae
Scientific Name	:	*Brachydanio rerio*
Native	:	Eastern India

This species is commercially known as zebra fish or zebra danio. It is native of eastern India. It is a very common aquarium fish and is said to be an excellent beginner's fish.

Identification

With narrow elongated body, they look beautiful under aquarium condition. They are robust, peaceful and sociable but sharp and turbulent. The body is extended

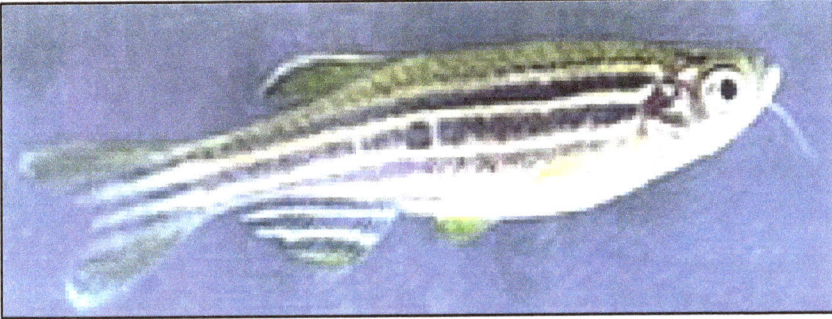

Figure 7.JD.4a: *Zebra danio*

Figure 7.JD.4b: *Brachydanio rerio*

with horizontal strips of dark colour white, gray and piroza, which even extend to the fins. It is very common aquarium fish and said to be an excellent beginner's fish. The body is extended with horizontal strips of dark colour white and grey, which ever extended to the fins. They prefer to be kept in schools of 5–8. Females are rounded shape and are usually larger in size, with the narrow elongated body. They look beautiful under aquarium conditions.

Full Length

The maximum size is 6 cm.

Water Quality

Temperature   :   24° C

pH   :   6.5 to 7.5 (neutral to slightly acidic)

Food and Feeding

They are omnivorous and will accept any type of food, but prefer live or frozen brine shrimp, small insect and worms. They also feed detritus.

Sex

Female have a rounded shape and are usually larger in size.

Breeding

Breeding pairs may be kept in a tank with fine leaved plants, among which the eggs are laid. They spawn in shallow water among marginal weeds and roots usually after heavy rain. Eggs are sticky and they exhibit cannibalism on the eggs. Hatching will be in 48 hours. The young ones will start free swimming on the fifth day.

Care should be taken to remove the parents after spawning to prevent eggs being eaten.

Aquarium Behaviour

They are active and peaceful. They prefer to be kept in schools of 5 to 8. They also prefer flowing water. They often jump out from the aquarium when frightened. They do well in planted tanks with plenty of free swimming space.

(JG) Gold Fish

Classification

Phylum	:	Chordata
Super Class	:	Pisces
Class	:	Osteichthyes
Sub-Class	:	Actinopterygii
Order	:	Cypriniformes
Family	:	Cyprinidae
Scientific Name	:	Cyprinidae

Identification

Basically there are two separate types of gold fish one denoted as scaled and other as scale less. Scale less goldfish posses' translucent nacreous scales in reality, which give them glowing colours such as blues and lavenders. The large scales on the goldfish may be pigmented or un-pigmented. These un-pigmented varieties give the scale less appearance. Aquarist has shifted from the term scale less to calico to define those goldfish lacking black pigment in their scales. From one common ancestor about 126 breeds of fancy gold fish have been developed. The best-known breeds of fancy goldfish are:

1) Shubunkin	2) Comet	3) Fantail
4) Veil tail	5) Telescope	6) Moor
7) Celestial	8) Lion head	9) Ornanda

10) Pearl Scale   11) Jikin   12) Ryukin
13) Tosakin

Common Gold Fish

It is almost olive green in colour but it soon darkens until it is almost black. Under good conditions it grows to a length of 9" or even more. It has a short body with a smoothly curved back that should show no sign of hump. The fins are not particularly large. The common gold fish is a stronger fish then most other varieties.

The Shubunkin (7.JG.1)

In the shape of the body the shubunkin is much the same as the common gold fish but rather slimmer. The fins are larger in Shubunkin and lobes of the fins are rounded. The shubunkin does not grow to a large size as the common gold fish. It has a blue body speckles with black.

The Comet (7.JG.2)

This has the normal goldfish colouring but has a long slim body with a caudal fin of about the same length as body. The fins on the under part of the body are long and pointed. The larger differently proportioned deeply forked tail fin makes comet goldfish one of the fastest swimming varieties.

The Fantail (7.JG.3)

This is one of the fancy varieties having a body that is the shape of a globe with two complete caudal fins and double anal fins. The body is often more oval than the ideal of rounded; its dorsal fin is less peaked. The fantail has a divided caudal fin of medium length.

The Veil Tail (7.JG.4)

The veil tail is another breed with both scaled and calico types and may have a swallow tail or broad tail. This breed is even more fanciful than fantail. The ideal body of veil tail is round. The magnificently flowing tail is quite twice the length of the body. The fork is entirely absent and the veil fall gracefully all the way down. Veil tail are seen with either normal or telescopic eyes. It is not as hardy as most other breeds and does not stand up well to a sudden change of temperature. The best temperature for gold fish is between 55° to 70°F.

The Pearl Scale (7.JG.5)

The name is derived from its dome shaped scales, with the outer edges of each individual scale being darker than the raised center. The scales are more convex than ordinary ones. It has a flat back, fat body and pointed small mouthed head. It is normally silver in colour distinguished by large red patches. It is interested to note that if one of the scales is lost, a normal scale will replace it, and the pearl scale will never grows for a second time.

The Moor (7.JG.6)

It has a dull black colour with no yellow or gold markings. The moor is not a particularly hardy fish and there is no calico variety. Two forward pointing but short

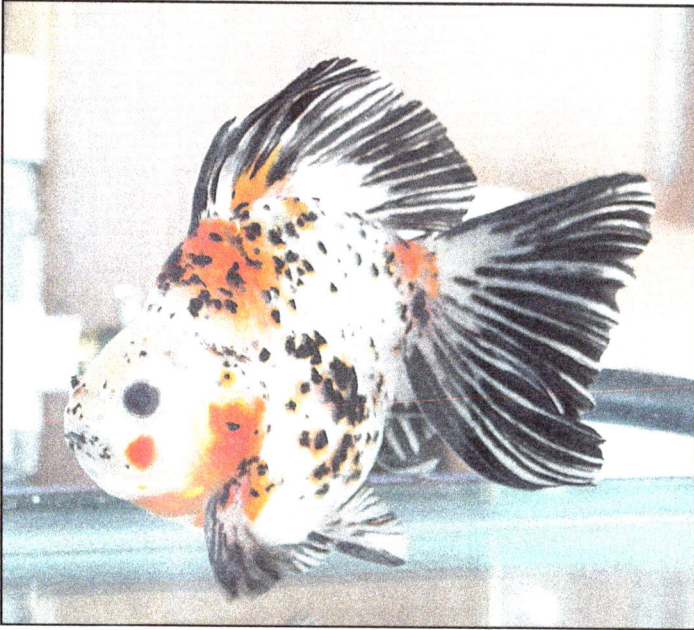

Figure 7.JG.1: Gold Fish Shubunkin

Figure 7.JG.1.1: Shubunkin Goldfish (Other variety)

sighted protruding telescopic eyes are found in the fish. They have telescopic eyes and have fins identical to the veil tails.

The Celestial (7.JG.7)

The projecting eyes are able to look upwards only. This obviously makes life far from easy for the fish, as it can feed only when its food is above it. Where it can be

Figure 7.JG.2: Comet Gold Fish

Figure 7.JG.3: Fantail Goldfish

Figure 7.JG.4: Veil Tail Goldfish

**Figure 7.JG.5: Pearl Scale
Gold fish**

**Figure 7.JG.6: Black Moor
Telescope**

**Figure 7.JG.7: Celestial
Bubble Eye Goldfish**

seen. It is not a hardy fish. The name is derived from the fact that pupils of its protruding eyes are always on top of the eyeball pointing upwards, *i.e.* heaven ward.

The Oranda (7.JG.8)

Wart like growth covers the whole head of the Oranda. It usually has a colouring of red and silver. The body shaping is similar to that the veital. The Oranda is reasonably hard. The Oranda gold fish are considered modification of the Lion head, having similar colouring but shapes more along the lines of veiltail possessing dorsal fin. The bodies of the best specimens are of even roundness, in the curved back as well as the full tommy, nicely filled flanks, a smoothly curving dorsal fin.

The Lion Head (7.JG.9)

The lion head has been developed from the Oranda. The warty head is even larger than that of the Oranda but the caudal fin is different, resembling that of the fantail. This variety has no dorsal fin. The lion head is less hardy than Oranda. The heads of these fish are enlarged and have an appearance. Which resembles a mass of the wart like growths. This distortion of the head reduces the flexibility of gill plates, so Lion head require well-aerated water to enable them to breathe without difficulty.

Figure 7.JG.8: Black Oranada

Figure 7.JG.9: Lion Head Goldfish

The Bubble Eye (.7.JG.10)

It is having two great big pouches under each eye, the whole eye looks as though it's dropped. Bubble eye has fluid filled bladders below their otherwise normal eyes.

Figure 7.JG.10: Teleoscopic Eyes

Figure 7.JG.11: Goldfish Jikin

Jikin (The Peacock tail Metallic group) (Figure 7.JG.11)

It is very old variety developed from Walkin having double caudal fin. Apart from the caudal fin, the fins and body are the Walkin, although slightly compressed vertically and some thicker in the region of the belly. The best specimens have a silver body with red lips and fins but perfect placement of the red is very rare. The distinguishing feature of this fish is the caudal fin, seen from behind is "X" shaped and the peduncle is broad. The axis of the caudal fin is almost perpendicular to the axis of the body.

Rhukin (Metallic Group)

It is Japan's second, most popular variety. In this fish the red deviation is seen towards the short, deep bodies' types. The body is short, deepest and moderately

Figure 7.JG.12: Tosakin Goldfish

compressed often with pronounced hump at a junction with the head. The fins are longer than those of the walkin, the caudal fin is forked and divided into two fins; the anal fins are also paired. The Rhukin is popular with professional fish breeders because of its hardiness and ease of management and the high percentage of good progeny with it produces. It is suitable for good pond and aquarium.

Tosakin (Figure 7.JG.12)

Tosakin probably a sport from the Rhukin, which is resembles. The main differences are a slightly shallower body and shorten fins together with a peculiarity of the caudal fin. The lower lobes and the caudal fins are greatly extended with up turned edges; the fins has the appearance of being reversed and spread pout in the direction of the head. Due to the difficulty in swimming they are unable to spawn naturally and must therefore, be stripped by hand. They are suitable for the aquarium only.

Black Moor Telescope

Common Name	:	Black Moor Telescope
Scientific Name	:	*Carassous auratus* Var. *bicaudatus*
Native	:	China, Japan, Asia.

Identification

Typical telescope fish are the greatly protruding eyes, projecting in the line of the optic axis symmetrically on both sides. Such eyes are the results of retina degeneration caused by hormones secreated by the thyroid gland Black telescope fish was developed in China. The body is relatively short. The caudal fin was originally short but in recent years has become longer through breeding.

Varieties of Gold Fish

Calico Gold

Common White Goldfish

Lion Head Gold Fish

Red Cap Goldfish

Red and White Goldfish

Red and White Butter fly tailed Goldfish

Full Length	:	36 cm
Food	:	Live, artificial and plant foods.
Sex Ratio	:	2 males : 1 female

Sexual Dimorphism

The male has a rush on the gill covers during spawning period. His violent chasing of the ripe female may also identify the male.

Water Quality for Breeding

Temperature	:	18–25° C
pH	:	7.0
Hardness	:	Max 2° dCH
Eggs	:	1 mm size, sticky type
Incubation period	:	72 hours/25°C. The hatching of the embryos may take several hours longer.

Breeding

Breeding tank should be specious stocked with plant thickets. This species may be kept in Mono-species tank. For breeding "Kaka Bandh" type arrangement may be provided. As eggs are sticky type eggs will stick to Kaka Bandh. After spawning Kaka Bandth should be removed, and to be placed in separate tank having same quality of water.

Gold fish have been bred in captivity for many centuries and breeds have taken advantage of many mutation of a hereditary character, which have occurred through the years. In this way many odd and colourful varieties have been produced including some with unusual scales, double tails and fins, no scales at all, missing fins, elongated or shortened back bones, telescopic eyes. Some of the better-known of the many existing varieties of the gold fish are described briefly:

Aquarium for Gold Fish

A straight side aquarium tank will keep them in a much healthier condition and also affords a much better view of its occupants. Five litters of water per 2.5 cm fish length is required to keep them in healthy condition in the aquarium. The maintenance of gold fishes under crowded conditions can be achieved only with frequent water changes.

The bottom of the aquarium should be covered with clean sand and gravel to depth of about 2.5 cm. The bottom can be planted with aquatic plants like *Cabomba*, *Vallisneria* and *Sagittaria* in the sand, preferably along the back of the tank so as to permit free movement of the fish in front. Aquarium plants draw most of their nourishment from the water and require an anchorage in the aquarium.

Food for Gold Fish

Most baby food cereals are excellent for gold fish. The fish may be fed sparingly every day in summer and every other day in winter. A pinch of food daily should be sufficient for two small gold fish.

Suggestions for Better Maintenance

Don't over crowd. Use rectangular aquarium rather than a glass bowl.

Don't change the water too often unless polluted; once in every 3 months should be sufficient. Occasionally, dip out 4 lits or so and replace with freshwater of the same temperature. Such water should be allowed to "age" for 24 hours before use. Avoid chlorinated water.

Never shift fish to water that is warmer or colder than that in the aquarium. The best temperature for gold fish is between 55° to 70° F.

Don't overfeed. Never place more food in the aquarium that can be consumed by the fish in about 20 minutes. Feed once only.

When the fish suck in air at the surface, it is a sign that they require more oxygen. Electric air pumps can help restore oxygen to an overcrowded tank.

At spawning time, remove all snails; you may never have baby gold fish without water plants in the pool or aquarium.

The simplest remedy for a sick gold fish is a mild salt bath for several days combined with fresh food.

Sustaining Capacity of Aquarium

A small aquarium cannot sustain much life. Generally it is as under 25 lit. aquarium can maintain two gold fish (each 5 cm long, exclusive of tails) in 80 lit. Aquarium (60 cm L × 30 cm W × 38 cm H) eight 5 cm fish of different varieties. When larger aquaria are used, the number of fishes and plants can be increased proportionately. If the fish are larger, their number should be reduced proportionately. Minimum requirements for tropical fish vary.

7 square centimeter of air space per full grown guppy, 20 square centimeter for full grown sward tail or platy, 48 square centimeter per medium sized barb and 15 square centimeter for large barb or 13 cm cichlid.

Water Quality Requirement for the Breeding of Gold Fish

Temperature	:	(1) Preferable temperature for growth and culture
		15.5 to 23.6 °C
		(2) For Breeding 20.00 to 23.0 °C
pH	:	7.5 to 8.0
Hardness	:	$CaCO_3$/liter: 50 to 70 ppm.
		Oxygen: Over 5 to 6 ppm.
		Free Ammonia: Less than 1 ppm.

Nitrite Nitrogen (NO_2): Less than 0.01 ppm

Nitrate Nitrogen (NO_3): Less than 20 ppm.

Photoperiod: 12 hours Sunlight.

Maturity

Size

Gold fish generally mature when they attain the size of 3 to 5 inches in length. But in the case of fancy varieties among them, they mature even at a lesser size.

Age

The attainment of age of maturity extends from 9 months to 2 years.

Appearance of the Anal Opening of Gold Fish

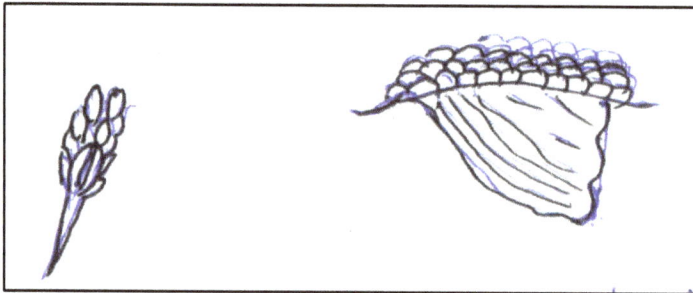

Male Profile   **Male Under View**

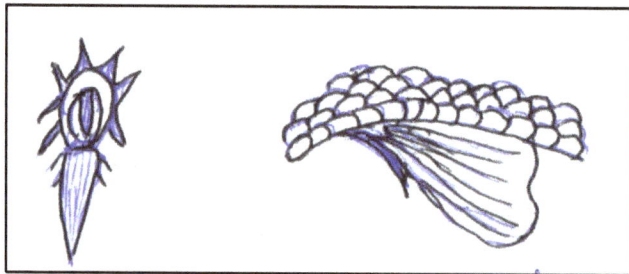

Female Profile   **Female Under View**

Showing Organs of Male and Female

Selection of Brooder

Male

In case of male, white bumps or tubercles develop on the operculum and pectoral fin, with the main ray of pectoral fin having ore pointed, lead ray of anal fin becoming thicker. Vent assumer a concave shape with small opening. Abdomen is seen to be smaller, slender, firm and with or without ridge. Further, the general body shape becomes slender, longer and symmetrical from the top.

Female

In female, few or no tubercle is found in the operculum region and pectoral fin. The leading ray of pectoral fin shows a thinner edge and other fins look rounder. Vent becomes convex and large, abdomen larger and fatty with no abdominal ridges. The general body shape of the female is influenced by fat, and it looks shorter and symmetrical from the top.

Sex Ratio

Generally male: female ratio is kept 2: 1 to ensure successful breeding.

Fecundity

Some of the females release more than 1000 eggs, the general fecundity of a healthy female is around 500–700 eggs depending upon the size. Natural spawning months are between May to September. If it is two years old 2000–3000 eggs and if it is four-year-old then 10,000 eggs with suitable breeding stock and correct feed. One spawn per week is seen.

It is advised to rest brooders for at least 3 months at lower temperature 12°C.

Survival

50 per cent survival of eggs in outer ponds is observed. 10 per cent mortality (*i.e.* 90 per cent survival) with well designed hatchery.

Breeding of Gold Fish

During on set of breeding season, the male chases and harasses the female. The female needs soft water and shallow area for breeding. When full with eggs the female release a pheromone to attract the male. Small gold fish will follow fish two or three times their size and rub against the larger fish. Soon some of the more active males will single out a female and drive her vigorously through the water. These are usual signs, which precede the actual spawning, but the more exotic type of fish may not go through these preliminaries.

Spawning commences early in the morning, usually as the first light reaches the tank, and continues to about noon. Once the spawning fishes have completed their activity, they will often turn around and eat the eggs themselves. With such cannibalistic inclinations, a young fish must be removed immediately.

A 50-liter capacity rectangular shaped aquarium tank can be used for breeding purpose. Eggs are generally released during night hours, after courtship and mating is over. Male discharges sperm simultaneous to release of eggs leading there to fertilization.

Fertilized eggs are transparent/grayish in colour and unfertilized eggs are milky white. Fungus grows very quickly on the surface of unfertilized eggs and very often the infection spreads to healthy fertilized eggs too. Considering this, it is desirable to remove the unfertilized eggs as soon as possible once the eggs laying is over.

Hatching

Fertilized eggs hatch in 4–7 days depending upon the water temperature. It is necessary to remove parents and other fishes, from the breeding tank if a separate hatching tank is not used. If this is not done the parents eat the hatchlings.

Feed to Hatchlings

The newly hatched young ones depend upon their yolk sac as food source for a couple of days. When the fry becomes free-swimming they can be preferably fed with *Artemia, Dephia, Monia* Tubifex worms, blood worms, influsoria and other phytoplankton.

Healthy fry of bright colour and good fin age should be reared in separate tank and fed with balanced diet for near about one month before transferring them to breeding tank. This process results in a greater success rate in breeding. Kept in a tank of appropriate dimension containing water of good quality, and fed on good food, they can grow up to 1–4 inches within six months.

Repetitive Breeding

The same parents can be deployed for breeding again after 3–4 weeks. For successful repetitive breeding, the fishes have to be kept in tanks containing water of good quality and fed with high quality nutritious diet.

Infection Control in Gold Fish

Many times the eggs and hatchlings are susceptible to external pathogens like *zoothamnium, cliates, protozoans,* fungus (*Saprolegnia and Aachlya*) etc. To prevent these infections, it is imperative to ensure good quality of water and provide nutritive food. If there is any pathogen infection, chemicals like acriflavin, Potassium permanganate, malachite green, formalin, Oxytetracycline, Nitro-flrazolidon can be used in proper does with utmost care.

Growth Calendar of Gold Fish

Time	Size
Newly hatched	4 mm
5 days old	5 mm (free swimming)
9 days old	7 mm
12 days old	9 mm
20 days old	12 mm
28 days old	15 mm
36 days old	18 mm
60 days old	20 mm
90 days old	22 mm
120 days old	24 mm
180 days old (6 months)	30 mm (Starts pairing)

Koi Carp

Classification

Phylum	:	Chordata
Sub-phylum	:	Vertebrata
Class	:	Actinopterygii
Sub-class	:	Neoptergii
Family	:	Cyprinidae
Scientific Name	:	*Cyprinus carpio* (Linnaeus 1758)
Native	:	Koi carp is the native of Japan.

Identification

The aquarium reared Koi is a colour variant of the common carp. Koi differs from gold fish in having prominent barbells of the corner of its mouth. Koi is available in a whole range of colour and shads.

Male have spawning tubercles while the female are more rounded when viewed from above. They are very peaceful and hardy fish.

Maturity

Koi become sexually matured by 2 to 4 years.

Water Quality

They prefer cold waters of temperature around 15–20° C and pH around 7.5.

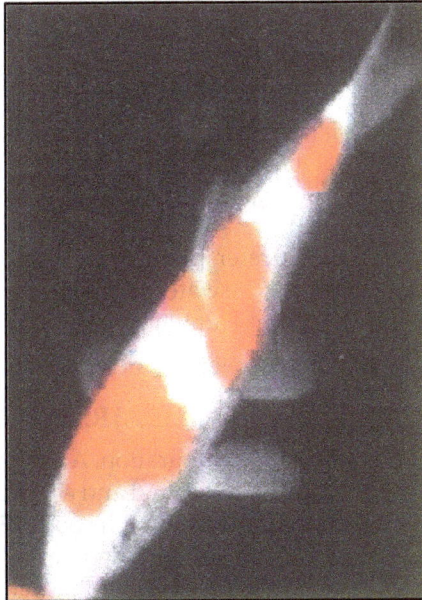

Figure 7.JG.13: Koi

Food

The food of Koi includes submerged plants and other benthic organisms. They readily accept artificial feed.

Breeding

Hand spawning techniques are also used to breed these fishes. The incubation of eggs is 3 to 5 days and fry starts feeding by the 10th day. At this period they may be fed with infusoria and egg yolk. The splendid breeds of Koi in the varities section were developed from selecting breeding.

(JR) Rainbow Fish

Common Name	:	Rainbow Fish
Scientific Name	:	*Malanotaenia* spp.
Native	:	Australia, Madagascar

There are about 50 species of *Malanotaenia* sp. known.

They are found in shallow water with abundant vegetation. They are mid water swimmers like barbs, lively and active. They require plenty of swimming space in the aquarium and can adopt wide range of water condition.

Water Quality Requirement

Temperature	:	22–28°C
pH	:	7 to 8.5
Hardness	:	Medium to hard.

Three Striped Rainbow or Jewel Rainbow (*M. trifasciata*) (Figure 7.JR.1)

They live in fairly hard, slightly acidic to alkaline water (pH 6.6 to 8) with temperature around 24–28°C. They prefer a planted aquarium with genital water movement. Several colour variations of the species exist in different regions; varying in adult size.

The Lake Kutubu Rainbow (*M. lacustris*)

They are peaceful fish and prefer planted aquaria with reasonable water movement and open space to swim. The water should be ideally be 22–24°C and pH range should be between 7–7.5 and medium hardness.

Males are colourful. They spawn over fine plants, preferably java moss.

Arfak Rainbow (*M. arfakensis*) (Figure 7.JR.2)

With silvery reflections and mauve colouration, Arfak rainbow looks brighter in captivity. There is a wide bluish mid-lateral band and a narrow yellow-orange stripe between each horizontal scale row on the sides of the body. Fins are translucent with bluish to mauve shading. This is a stream dwelling variety found mainly around sub-surface vegetation submerged logs or branches in small tributary streams. The maximum size is 10 cm. Males are brightly coloured and larger and deeper bodied than female.

Figure 7.JR.1: Bottom: Three Strip *M. trifasciata*
Figure 7.JR.2: Top: *M. arfakensis*

Boeseman's Rainbow (*M. boesemani*)

With completely different colour pattern, the Boeseman's rainbow is distinguishable from other rainbow fishes. The head and front portion of the body are brilliant bluish gray, sometimes almost blackish with the fins and posterior half of the body largely bright orange-red. The wild colouration can fade in aquarium conditions. They are found in shallow water with abundant vegetation. Males are easily distinguishable with their brighter colour and elongated dorsal fin rays.

(JR) Rasbora

Family   :   *Cyprinidae*

Native

They are inhabitants of the freshwater streams of Malaysia, Jawa, Sumatra and Thailand.

Main Species

1. *Rasbora eithoveni* (Brilliant rasbora)
2. *R. maculate* (Pygmy rasbora)
3. *R. brorapetensis* (Red tailed rasbora)
4. *R. heteromorpha* (hari-quin rasbora)
5. *R. pauiperforata*
6. *R. trilineata* (Scissortail rasbora)
7. *R. kalochroma* (Clown rasbora)
8. *R. elegans* (Elegant rasbora)
9. *R. vateriflories* (Fire cey-ceylonese rasbora)
10. *R. caudimaculata* (Red scissor tail rasbora)
11. *R. steineri* (Golden striped rasbora)

12. *R. teaniata* (Black striped rasbora)
13. *R. dorsiocellata* (Eye spot rasbora)
14. *R. parlueiosoma libosa*

Identification

R. maculate (Pygmy Rasbora) (Figure 7.JR.3)

It is the smallest among the rasboras and it should be kept with compatable sized fishes. It prefers to have dense foliage and dim lighting.

R. brorapetensis (Red Tailed Rasbora)

Rasbora is very popular fish among hobbyists. They are active swimmers. They are with slim pale yellow body and distinctive dark and gold bands extending from gill

Figure 7.JR.3: *Rasbora maculate*

cover to the end of the caudal peduncle. A dark line runs along the base of the anal fin. The caudal fin is red in matured fish. Males are smaller than female.

R. heteromorpha (Figure 7.JR.4)

It is very popular fish. This species differs from the relatives with unusual shapes, marking and colouration. Its body is reddish copper colour, which is accented by a triangular shaped striking blackish–blue marking covering the rear half of the body. The marking is wide in the center of the body and becomes narrower as it extends backwards to where the caudal fin begins. The dorsal fin is red in colour with yellow

Figure 7.JR.4: *Rasbora heteromorpha*

tip. The upper and lower tips of the caudal fin are bright red with inside rays pale yellow. The body is silver coloured and the sides range from pink to copper.

Environment Preference

They prefer an environment with dense vegetation open area for swimming, a dark substrate and dim lighting.

Figure 7.JR.5: *Rasbora pauiperforata*

Figure 7.JR.6: *Rasbora trilineata*

Figure 7.JR.7: *Rasbora parluciosoma libosa*

Water Quality Requirement

Temperature	:	22–27°C
Hardness	:	1.5 to 2.5 dH
pH	:	Around 6 to 6.6

Breeding

They are egg layers. Spawning pairs prefer broad leafed plants such as *Cryptocornye* or *Aponogeton.*

Hatching

The larvae hatch within 48 hours.

Feeding to Fry

The larvae feed on yolk sac initially, after which they will accept tiny live food such as rotifers, Cyclops and artemia nauplii.

(K) Panchax

Order	:	*Cyprinidontiformes*
Family	:	Cyprinidontif

Scientific Name of Different Species

1. *Apolochelius dayi* (Green panchax)
2. *A. panchax* (Blue panchax)
3. *A. lineatus* (Striped panchax)
4. *A. blockii* (Rainbow panchax)

They are small surface swimming fishes. They are native of the Asian Continent distributed in Cambodia, India, Indonesia, Malaysia, Myanmar, Sri Lanka, Thailand. They are not found in China, Phillippines. There are eight wild species known. However, above four species are very popular in the context of aquarium interest. *A. panchax* has two species *viz. A. panchas panchax* commonly found in Bangladesh, Nepal, India and Malayan peninsula and *A. panchax siamensiis* which is found only in Thailand.

General Size

4.5 to 60 cm.

Habitat

They have an elongated and robust body with convex back at the insertion of the dorsal fin and the broad mouth and are found in the wild is a streams, rivers, low lying paddy fields, swamps and brackish waters.

Type of Breeding

They are egg layers. They lay their eggs on plants.

Water Quality

Temperature	:	24–27°C
pH	:	6.7 to 7.3

Details of Species

(1) *Aplochelius lineatus* (Stripped Panchax) (Figure 7.K.1)

They are native of India and Shri Lanka. They are found in three different colour variations–stripped, gold and red. The body is elongated and robust with a broad mouth, a characteristic of the genus. The dorsal fin is positioned above the hind part of the large anal fin which in the rear end is pointed in the male and rounded in female. The second pectoral fin ray is elongated, reaching back to the start of the anal fin in the female, considerably further in the male. In the caudal fin, the central rays are considerably larger than the others particularly in the adult males.

Maturity

They can attain a maximum size of 4.5m inches but will mature sexually at 3 inches long. Male and female are about the same size.

Ideal Water Quality

Temperature	:	25.5 °C
pH	:	6.8 to 7.2
Hardness	:	Around 7

They prefer conditioned water and regular changes of water will keep the fish healthy.

Feed

They eat floating feeds.

Figure 7.K.1: *Aplochelius Aplochelius lineatus*
(Stripped Panchax)

(2) *Aplochelius blockii* **(Raibow Panchax) (Figure 7.K.2)**

They are native of Indian peninsula especially the southern States. It is found in the paddy fields, small streams and drains. This is very active but not aggressive and most popular fish among all panchax. They breed profusely under favorable water conditions.

Figure 7.K.2: *Aplochelius blockii* **(Raibow Panchax)**

(3) *Aplochelius panchax* **(Blue Panchax) (Figure 7.K.3)**

They are native of India, Indonesia with slender body and rounded tail. It is better known as blue panchax, through by sight it appears green. However, with right lighting in the aquarium they will show a bluish shine. The females are pale green and have clear fins and black dot at the base of dorsal fin. The body is gray above with a shiny white spot on top of the head. This is an adaptable fish easily recognized by their shiny white spot.

Figure 7.K.3: *Aplochelius panchax* **(Blue Panchax)**

(4) *A. dayi* **(Green Panchax) (Figure 7.K.4)**

They are native of Sri Lanka. They are better known as Ceylon Kill-fish among the hobbyists. It is found in brackish water environment mostly in the mangrove

Figure 7.K.4: *A. dayi* **(Green Panchax)**

areas, rural streams and drains. They prefer moderately hard and alkaline water with temperature range from 24–27° C. It is very easy species to keep and breed.

Feed

It eats all kind of feed.

Habitat

They are found in the mangrove streams in small groups. It can tolerate and thrive in both freshwater and brackish water.

Water Quality

Temperature	:	26°C
pH	:	6.7 to 7.3

Water conditions are not extremely important as long as the water is not too hard.

Breeding

They will show if kept in a pair or small group in an aquarium tank containing a float spawning. The species are quite prolific breeders and medium sized eggs are picked from spawning mops and placed in plastic container to hatch within two weeks.

Feed of Fry

The fry are large enough to eat live baby brine shrimp and micro worms.

(L) Spiny Eel

Family	:	Mestacembelidae.
Scientific Name	:	*Macrognathus aculeathus*

Distribution

They are found in India, Pakistan, Sri Lanka, Thailand, Malaysia, Vietnam and Indonesia.

Identification

Spiny eels are not true eels but are called so due to their similar appearance to eels. The body is long and eel-like with a long fleshy snout and rounded caudal fin that is separated from the dorsal and anal fins. The body colour is brownish to yellowish and marked with dark bands or blotches. Due to its beautiful and slender body structure, colourful and playful behaviour they have great demand in the ornamental fish trade. Spiny eels are very susceptible to fungal and bacterial infections. Hence, special care should be given if it gets injured to prevent infection.

Macrognathus aculeatus and *Mastacembalus armatus* are of much importance as far as the ornamental fish market is concerned.

Figure 7.L.1: *Macrognathus aculeatus*

Colour

It has bare light brown body with round four ringed spot on the top of the long dorsal fin. Both the caudal and dorsal fins have several fine streaks.

Maximum Size

25 to 30 cm when fully grown.

Feeding Habit

They have nocturnal feeding habits, preferring to hide by burying themselves in the substrate or hiding under rocks during the day and coming out at night or early morning to feed. They are carnivorous and young ones feed on live food such as brine shrimp and variety of worms especially black worms when they become larger, they will eat small fish.

Water Quality

pH	:	Neutral to slightly alkaline having pH of 7.2
Temperature	:	24–26°C

Breeding

They can be bred in captivity. It is difficult to determine the sex when they are young. Females are slightly larger than male of the same age of female. During the breeding stage, the female develops a swollen abdomen with a greenish tinge and clear anal papillae.

Eggs and Hatching

Females lay 800 to 1000 transparent and large eggs. The eggs will hatch within three days.

Food to Fry

The young ones are fed on *Cyclops* nauplii

Mastacembelus armatus (Tire Track Eel or Zigzag Eel or Fire Spotted Eel)

Distribution

It is found streams and rivers with sand pebbles or substrate of Asia.

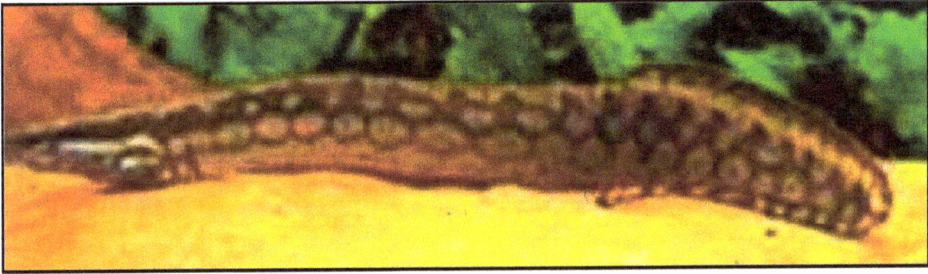

Figure 7.L.2: *Mastacembalus armatus*

Identification

They have a row of spines along the back. The body is elongated with golden brown colour and has two rows of darker brown longitudinal zigzag lines: more of less connected to form reticulated pattern. Dorsal, anal and caudal fins form one complete until encircling the rear.

Maximum Size

Maximum attainable size is 90 cm.

Habit

It is an interesting fish and would be a good addition to a large aquarium. They generally hide in the substrate or under rocks during the day and come out only at night and early morning to feed. Hence, it would be ideal to keep enough hiding places in the aquarium.

Water Quality

pH	:	6.4 to 7.5
Temperature	:	22–28 °C

Food

They have a small mouth and feed on worms and mosquito larvae and may also accept some diet foods such as pellets and freez-dried blood worm.

Breeding

'They are not known to breed in the aquarium or in captivity. The female will have a larger stomach when ready to spawn.

(M) Badi

Scientific Name of Sub Species: Nandidae

1. *Badis badis badis* (Figure 7.M.1)
2. *B. badis siamensis* (Figure 7.M.2)
3. *B. badis burmanicus* (Figure 7.M.3)

Native

India, Thailand and Myanmar.

It is better known as the dwarf Chameleon fish as it changes its colour quite often depending on mood and environment, specially during courtship and breeding. This fish have been around the hobby for years. They are very pretty like the dwarf cichlids.

Adult Size

6 to 8 cm.

Figure 7.M.1: *Badis badis badis*

Figure 7.M.2: *Badis badis salnensis*

Figure 7.M.3: *Badis badis burmanicus*

Colour

They are mildly colourful with series of orange and blue, with blue spots. However, when guarding a nest or group of young ones, the male becomes jet black with bright blue highlights in the fins. When frightened, they fade to beige colour with a dark spots on the caudal peduncle, when the dominant male is feeling blood, it turns dark with black strips running vertically down the sides.

Food and Feeding

They are almost carnivorous and feed on tubifex, white worm, shredded earth worm and other meaty foods and will not take any flake food or vegetable matter.

Behaviour

They are peaceful, slow moving and excellent fish, which go well in community tank.

Other Requirement

1. Ideal water temperature: 26°C
2. pH: 7.0
3. They required lots of plants and usually driftwood is used for giving shelter.

Badis badis badis (Figure 7.M.1)

It is a native of the Indian sub continent especially from the northeastern side. The maximum attainable size of the adult fish is 8 cm. Male are ventrally concave and much more colourful and care for their young ones. They can be kept in pairs or small groups in larger tanks and feed on live feeds and some frozen worms. They need lots of plants and prefer hiding places in the aquarium.

Breeding

The breeding pattern is almost similar to Cichlids. The male is distinguished by a similar appearance and more intense colouration. Males and Females should be separated and conditioned for a week on live feeds to get best results during breeding. The water should be slightly acidic to neutral (pH 6.8), the temperature 26°C. After preliminary courtship, which coincides with vigorous chasing by the male, the female will eventually be down in to a shade to lay eggs, which will be fertilized by the male. The parents should be removed to avoid mortality of the young ones.

Hatching

Hatching occurs within 72–80 hours. The young ones will be gradually fed on infusoria brine shrimp, daphinia and finally small white worms or chopped tubifex.

(N) Molly

Black Molly or Atlantic Molly (Figure 7.N.1)

Classification

Order	:	Cyprinodontiformes
Family	:	Poicilidae
Scientific Name	:	*Poecilia sphenops* (Cuvier and Valenciennes 1846)
Synonyms	:	*Gambusia modesta, Platypoecilus mentalis, P. spilonotus, P. tropicus*
Native	:	Maxico to Colombia in fresh as well as brackish waters. Central to North–Eastern South America.

Other Important Species

1. *Poecilia formosa* (Amazon molly)
2. *P. tatipinna* (Sailfin molly)
3. *P. maxicana* (Short fined molly)

Identification

It contain jet black colour. The male is smaller more slander has a gonopodium and his dorsal fin is larger and tinted orange. It raises dorsal fin to attract the female during spawning period. Females are larger than male.

Figure 7.N.1: Black Molly

Full Length

Male: 5 cm; Female: 7 to 11 cm

Food

Live-artificial and plant foods.

Sexual Maturity

12 to 16 weeks to mature.

Fertilization

Unlike other fishes, in the case of live bearers, fertilization is internal. Male counterpart transfers milt by an organ known as "gonopodium", which is a modified anal fin. There have been reservations, whether or not the gonopodium is actually inserted in to female body through vent. At one it was believed that the gonopodium was a hollow tube that transferred mild into the female or at least towards the female genital opening. Study with the help of electron microscope has revealed that in most livebearers, the gonopodium is gooved along its upper surface, and the milt passes through the semi-circular grooved tube of gonopodium into the female body.

Breeding

For mass breeding a tank size 100 × 100 × 60 cm is ideal. A perforated basket (waste bin) could be provided in one corner of the tank wrapped with fibrous plastic following filaments, for the female to drop the young ones. Soon after the birth the young ones escape from their mother and enter into the perforated basket and later the young one can be collected from the basket and placed in a separate tank for further rearing. Generally they give 50 to 100 fry per breeding.

Interval Between Births

4–8 weeks.

Water Quality

		Newly born babies	Adults
Temperature	:	22–24° C	24–27° C
pH	:	7.0–7.5	7.0–7.5

Add 1 teaspoonful of NaCl for every 10 lit of water.

Feeding to Fry

Fine live as well artificial food, after 14 days vegetable foods also can be given.

Feeding to Adults

They consume wide range of vegetable based food. They also consume live as well dried food.

Figure 7.N.2: Marbel Molly (*Poicelia latipinna*)

Figure 7.N.3: Golden Ballon (*Xiphophotus variatus*)

White Molly

Family	:	Poeciliidae
Scientific Name	:	*Poecillia sphenops*

Identification

This variety, with a fully white body, was developed from the marbled molly. White molly is now commonly bred in aquarium world over. All other characters and breeding method are just like Black molly. They are livebearers.

Spotted Molly (Figure 7.N.6)

Scientific Name	:	*Poecillia latipinna* (Le Sueur 1821)

Identification

Deep black body of this fish bears scattered white sports, which makes them more attractive and earned the name "Spotted molly" smaller and thinner male members of this fish grows to a maximum length of 10 cm. They carry a prominent gonopodium and a high and wide dorsal fin.

Breeding

Breeding is similar to all molly. Fish breeders have been successful in developing large varieties of these fishes.

Food

These peaceful community fishes prefer algal matter as their main food. However, they accept artificial prepared food also.

Sward Tail (Figure 7.N.4)

Common Name   :   Sward Tail

Classification

Order   :   Cyrinodontiformes

Family   :   Poeciliidae

Scientific Name   :   *Xiphophorus helleri*

Identification

It is well known orange and white colour aquarium fish. The male has a gonopodium, the last lower ray of caudal fin extended in to a "Sword" like projection. Hence it is known as sword tail fish.

Full Length

Male: 10 cm and Female: 12 cm.

Food

Live, artificial and plant foods.

Water Quality

Temperature   :   22°–25° C.

pH   :   7.0 to 7.5 Freshwater is preferred at the time of spawning. (Add one teaspoon NaCl for every 10 lit of water)

Figure 7.N.4: Sward Tail (*Xiphophorus helleri*)

Type of Breeding

They are live bearers.

Fecundity

Mature female may produce about 50–60 young. Interval between births 4–6 weeks. They are highly cannibalistic on fry.

Sex Ratio

1 male: 5 female

Breeding Tank

With a capacity of 10 liters per 1 female + breeding trap.

Feeding to Fry

Fine live and artificial foods, after 14 days plant food may be given.

Rearing of Fry

The number of fry in one liter may be any where between several dozen to 200 or more. The fry should be reared in large, low tanks with water column 10 to 15 cm high. In commercial breeding use is made of large breeding traps holding several dozen females. It is important to reduce the number of fish in the tank in time and provide a continuous supply of freshwater (through flow tanks) and intensive feeding 3 to 4 times a day. From the end of June/July till the beginning of September the young fry may be kept in garden pools.

Sexual Transformation

Each immature fish could develop into either male or female depending on which reproductive organs develop first. If the ovaries develop first, these will secrete female hormones (estrogen) and the fish will develop ion to female. In case testes develop first they turn out to be males because of androgen secretion. Later on life, however, a female fish can turn into a functional male. Female to male change is common, but the reverse transformation is very rate. Sometime, the external factors such as pH can affect the sexual development of some fishes. It has observed that a low pH of 5–6 helps in developing more of males in the broods of sword tail and pH value more than 7.0 results in more of females. Sexual transformation in *X.helleri* has been described many times. Older females that have produced several litters may be transformed in to males. However, reverse transformation of males into females has not been known to occur in this fish.

Platy (Figure 7.N.5)

Common Name	:	Platy
Family	:	Poeciliidae
Scientific Name	:	*Xiphophorus maculatus*
Synonyms	:	*Platypoecilus maculatus, P. nigra, P. pulchra, P. rubra, Poecilia maculata*

Figure 7.N.5: Platy (M and F)

Identification

This is world famous species has become rightly popular due to attractive rich colour. They are having variable colouration even in the wild. The male can be identified with elongated dorsal fin and a gonopodium. Generally they are orange in colour with black caudal fin. They can be placed in community tank. They are live bearer variety.

Full Length

Male: 3.5 cm; Female: 6.0 cm

Food

Live, artificial and plant foods.

Sex Ratio

1 male: 3 female.

Breeding

Breeding tank with 10 lit capacity with breeding trap would be arranged water temperature 20–25°C, pH 7.0 to 7.5.

You have to add one teaspoonful NaCl for every 10 liters of water. Breeding is internal. Single female gives birth to 30 to 100 babies after 20–25 days of mating. This species readily and prolifically cross breed with *Xiphophorus hellieri*; the male hybrid offspring have only a short sward extension.

Feeding to Fry

Fine live as well as artificial foods.

(O) Guppy

Common Name	:	Guppy or Million Fish
Classification	:	Guppy
Phylum	:	Chordata
Super Class	:	Pisces
Class	:	Osteichthyes
Sub Class	:	Actinopterygii
Order	:	Cyprinodontiformes
Family	:	Poecilliidae
Scientific Name	:	*Poecilia reticulate*

Identification

Guppies are small fishes with a vast variety of caudal fin evolved through crossing and selection. The dorsal fin is irregular. The eyes are large and mouth is up turned, suitably adapted to feed on the water surface. The colours of guppies show extreme variations. The common colours are red, blue, green, gray and yellow with spots. The maximum size is 8 cm. The females are larger in size than the males. The males are provided with gonopodium. The males are beautifully coloured than females. The male has a swollen second ray on the pelvic fin, rarely resembles other males in appearance.

Figure 7.O.1: *Poecilia reticulate*

Full Length

Male: 6 cm; Female: 8 cm.

Food

Live and artificial foods, regularly augmented by plant food.

Tank

They can stay in community tank. However, they are kept in mono species in commercial breeding centers.

Sex Ratio

1 Male : 3 Females.

Water Quality

Temperature	:	22–24° C
pH	:	7.0 to 8.0

Add one teaspoon NaCl for every 10 liters of water.

Breeding

They are livebearers. They should be kept in breeding tank with capacity of 3 to 10 liter and breeding trap should be provided. The male's iridescent colours attract the female Guppy. Courtship may begin with numerous behavioral patterns, two of which are known as "Following" the female and "Biting" at her genital region. In some of these actions the male display his median fins. If he succeeds in swimming the female attention, he responds by showing her swimming. Then the, male commences series of sigmoid display, lending his body in to S shaped curve before the female, so that light reflects his iridescent colours. It is during this final performance that he may make an attempt to copulate, maneuvering into such a position that he can insert the tip of his gonopodium into the female's genital pore. The pair swim in a high circle, with the female flanked by the male, while she archer her hack to expose the genital region, the male transfer bundles of sperms which break up inside the female and migrate to storage pouch near the ovary, where they are retained until required. The eggs mature in the ovary and fertilized before released, so that fertilization occurs prior to ovulation. Guppies can develop several young at different stages simultaneously, a phenomenon known as suerfoetation. Since it is cannibalistic in nature, to save the young ones, a "Breeding trap" is used.

Fecundity

Females give birth to as many as 250 young at the intervals between 4–6 weeks.

Feeding of Fry

Finely sifted zooplankton, brine shrimp nauplii, may be given, after 14 days occasional vegetable food can also be given to fry. From the small tank pour off fry of the same stain in to larger tank with a water level of about 10 cm and raise this level to as the young fish grow.

Varieties of Guppy

Guppy (Male)

Black Guppy

Delicate Variegated Guppy

Green Variegated Guppy

Blue Variegated Guppy

Red Snakeskin Guppy

Green Snakeskin Guppy

Yellow Snakeskin Guppy

Tuxedo Guppy

Tuxedo Cobra (Dragon Head) Guppy

Red Tail Guppy

Blond Red Tail Guppy

Blue Tail Guppy

Pineapple Guppy

Double Sword Guppy

Electric Blue Guppy

Leopard Guppy

Lyretail Multicolor Guppy

Better Growth

As soon as is possible to identify the sex of the fry separate the males from the female. The later the males attain sexual maturity the greater is the probability that they will reach the largest possible size. Because the generations follow quickly one upon the other, in time, with constant in breeding, there are degenerative changes in fish, particularly a decrease in their size. It is therefore, advisable to obtain brood fish of the same strain but from different unrelated groups.

Mosquito Fish

Scientific Name : *Gambusia affinis affinis*

Identification

D-67, A-8–10, LL 29–30.

This species is readily distinguished from Guppy by its spotted caudal fin. It is famous for its preferred diet of mosquito larvae. It is slightly larger than guppy and eats its own weight of larvae daily. The female is fairly dull olive-green with scattered black spots. The colour of male is very variable, being basically yellowish with a pale belly and often with a lot of black marking, especially on dorsal fin tail and under eyes. The first 4–6 rays of the pectoral fin are thickened and curved upward on their distal half (the half furthest from the body) to form bow or notch.

Characteristics of Gambusia

Uses aid stabilization of the gonopodium in its anterior swing during copulation.

Full Length

Male: 3.5 to 4 cm; Female: 6.5 cm

Food

Largely carnivorous and eat mosquito larvae, but will accept virtually any other food offered larvae as well as artificial foods.

Tank

Mono species keeping tank.

Breeding Tank

Breeding tank with a capacity of 10 liter with breeding trap may be provided.

Sexual Maturity

Gambusia becomes sexually mature at just a couple of months old. Probably because of pressure from predation, since they are very vulnerable living at the water surface.

Figure 7.0.2: *Gambusia affinis affinis*

Sexual Dimorphism

The male is smaller, slender and has a long gonopodium. Pregnant females bear a black blotch on the abdomen and produce up to 80 juveniles.

Tolerance Limit

It has been quite well known for its voracity and its tolerance of wide range of conditions from freshwater to brackish swamps varying between 10° C and 30° C.

Breeding Season

Breeding season is from April to October.

Sex Ratio

1 male : 2 female.

Water Quality for Breeding

Temperature	:	22–28° CV
pH	:	7.0 to 8.0

Add 1 teaspoon NaCl for every 10 liter of freshwater.

Type of Breeding

They are livebearer. Interval between two births remains 4–6 weeks.

Gestation Period	:	30 days
Production of Fry	:	10–100 fry

Feeding to fry: Brine shrimp nauplii, finely sifted zooplankton, artificial fry foods etc. The young fish grow quickly.

Temperament

It is unsuitable for inclusion in community tanks because of its aggressive nature.

Basic Care

Hungry mother eats their fry. Therefore, immediately female should be removed from tank or breeding basket may be provided so that fry can escape.

(P) Oscar

Scientific Name	:	*Astronotus ocellatus* (Cuvier 1829)

Native

Native of this species is South America. However, it is popular all over world.

Identification

The fish is olive green in colour with a few fiery red markings on the body. The dark caudal fin has an orange ring in its base. They should be kept in mono species tanks with suitable hiding places.

Figure 7.P.1: Oscar

Full Length

It grows to 30 cm in size.

Water Quality

They prefer soft water of pH around 6.5 and Temperature ranging between 22 to 27°C.

Breeding

The Oscar should be allowed to select their own mate. For this purpose, 10 to 12 fishes are to be reared together. As soon as they become mature, they select their own mate. A pair can be distinguished by its habits of cleaning rocks together and chasing other fishes in the aquarium. It is not easy to differentiate the sexes. Oscar spawns on smooth round rocks. Hence the aquarium tank should have big rocks at the bottom.

The barrel shaped eggs attached to the rocks, hatch in 3 days and fry are moved to a pit nearby, prepared by the parents. Some Oscar tend to consume their own eggs, in the such cases, it is advisable to transfer the eggs using hose in to an aquarium with similar water conditions. Few drops of methylene blue should be added apart from providing profuse aeration. They starts exogenous nutrition in four to five days. They should then be fed on a diet of Artemia nauplii. As they grow up, they can be fed on chopped tubifex worms. The juveniles are quite different from adults in appearance. The body is dark brown to black in colour with numerous patches on it.

It is advisable to provide hiding place like piece of pipe, pot etc, in the aquarium.

(Q) Puffer Fish

Common Name	:	Puffer fish, Blow fish, Balloon fish.
Family	:	*Tetraodontidae*

Introduction

There are over 150 species are known from freshwater and brackish water environment. Only small numbers of species are of interest to the aquarium trade.

Important brackish water and freshwater species are listed as under:

Important Brackishwater Species

1. *Tetraodon mbu*
2. *T. nigroviridis*
3. *T. fluviatilis*
4. *T. biocellatus*
5. *T. lineatus*
6. *T. leiurus*

Important Freshwater Species

1. *Carinotetraodon lorteti*
2. *C. salivator*
3. *C. somphongsi*
4. *Monotetrus travancoricus*
5. *Chonerthinos amabilis*
6. *C. nefastus*
7. *C. monlestus*
8. *C. remotus*
9. *C. asellus*
10. *Colomesus ansellu*
11. *C. psittacus.*

Distribution

They are having world wide distribution.

Identification

Puffers get their common name from their unique defensive ability to inflate their bodies when threatened by the predators. Due to their ability to puff up, they seem to hover about their tank, looking distinctly like a golf ball, with bulky body contour. They are scale-less and most of them have spines on their bodies, which through are not always obvious when the fish is not inflated. Unlike other fishes, they do not have pelvic fins. Instead of using the caudal fin as the main organ to swim, Puffer swims with their transparent pectoral fin, gaining guidance from dorsal and anal fins. The eyes of puffer are able to move independently of each other, giving binocular vision.

General Behaviour and Habit

They are considered to be a long living fish even in captivity and relatively simple to care for. They can easily acclimatize themselves to fluctuation in water conditions as long as it is gradual transition. Puffers are quite susceptible to inflections and parasitic disease. They are known for nipping at the fins of other puffers and species of fish, which can be minimized by keeping from all fed. However, freshwater dwarf puffers do less nipping compared to other types of puffers. Small varieties of puffer travels in to freshwater to breed although are actually of marine origin.

General Feeding Habit

Though they are carnivorous, they will quickly adapt to prepared food. However, they prefer small molluscs, shrimps and krill with their shell on to wear down their even-growing teeth.

Details of Some of the Species

(1) *Monotetrus travancoricus*

This is popularly known as Malbar or dwarf puffer fish. It is native of India and considered the smallest puffer in the trade. With a maximum size of 2.5 cm, this fish is the current dwelling of the freshwater puffer hobbyists.

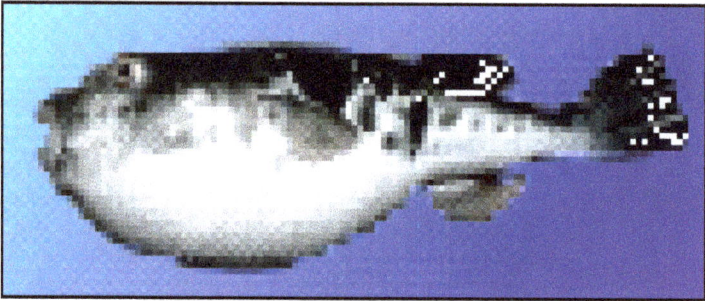

Figure 7.Q.1: *Monotetrus travancoricus*

Feed

They feed on small meaty food and leave plants alone.

(2) *Tetraodon fluviatilus*

It is known as the Ceylon puffer. It is native of Sri Lanka, India, Bangladesh, Myanmar and Borneo. They are common puffer in the aquarium fish trade and are aggressive fin and scale nippers as they become adults. For breeding sex ratio may be kept 1: 1.

Figure 7.Q.2: *Tetradon fluviatilus*

Feed

They feed on small crustaceans, worms, molluscs, algae and detritus in the wild.

Eggs

Incubation period if 4–5 days.

(3) *Tetraodon nigroviridis*

They are known as the green spotted puffer due to its characteristic emerald green colour, which makes a nice contrast to the dark spots. They are native of tropical eastern Asia, from the coastal regions of Indochina and Philippines to India. It is found in fresh and brackish water and is a common puffer in the aquarium trade. Their maximum size is about 15 cm. They are very aggressive and often kept alone. The water condition is medium to medium hard with pH 8 to 8.5 dH 9.0 and temperature 24° to 28°C.

Figure 7.Q.3: *Tetradon nigroviridis*

(4) *Tetraodon mbu*

They are commonly known as giant freshwater puffer or mbu puffer. They are native of Africa, widely distributed in Lake Tanganyika and the river Congo basin.

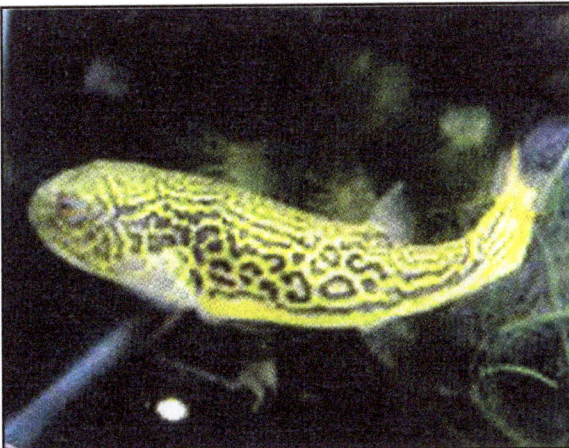

Figure 7.Q.4: *Tetradon mbu*

This is one of the largest freshwater puffers. The attainable maximum size is more than 75 cm. Water conditions are medium pH 7.0 to 7.2 with temperature of 24 to 26°C.

(5) *Tetraodon biocellatus*

It is commonly known as the "Figure 8 "puffer or eye spot puffer. They are native of Indochina, Malaysia and Indonesia. They are of brackish water origin and are aggressive. The best water parameters are pH 6.5 to 7.5 snf dH 5 to 12. The common size is little over 5 cm.

(6) *Tetraodon leiurus*

They are commonly known as the Target puffer or twin spot puffer.

They are native of Asia. It can withstand water conditions such as freshwater to brackish with pH 7.0 dH 12.0. It is very aggressive, being a fin nipper.

(7) *Tetraodon lineatus*

They are commonly known as Fahaka puffer or Nile puffer. They are native of Nile, Chand basin, Niger, Volta, Gambia, Geba and Senegal rivers in Southern Africa. They are also known as lined puffer. They can be acclimatized from fresh to brackish water conditions with a pH of 7.0 and dH 10.0 maximum size is 35 cm.

(8) *Carinotetraodon somphongsi* (Freshwater)

Synonyms

1. *Tetraodon somphongs*
2. *Carinotetraodon chlupatyi*

They are native of Thailand in freshwater only. Full length is 6.5 cm. The dorsal fin rays of the male are rust-red; the caudal fin is white edged with a darker colour, and the body colouring is grayish. The female is paler hue with more distinct markings on the upper half of the body. They should be kept in mono species tank.

Breeding

Sex ratio may be kept 1: 1. They lay eggs of 0.5 mm, whitish transparent, very sticky. Incubation period is 60–m 72 hours. Sufficient hiding possibilities and thick plants should be provided. They prefer 26–28°C pH 6.5 to 7.0.

Feed

They feed on live foods, like gastropods, earthworms, tubifes worms, fry feed Cyclops nauplii.

(R) Mouth Brooder

Scientific Name : *Tilapia mosambica*

Identification

It is an important fish found in fresh as well as brackish waters. This is an aggressive fish towards other members of the same species. The dominant male in the social hierarchy is intensely colored.

Figure 7.R.1: *Tilapia mosambica*

Full Length

40 cm (The fish are already sexually mature when 10 cm long)

Food

Live, artificial and plant foods.

Sex Ratio

1 Male : 1 Female

Colour

During the spawning period the male is deep black, the female gray-green.

Breeding

The female deposits several hundred eggs in the pit, but simultaneously gathers them up into her mouth together with milt. During their development in the mouth the female does not take food and grows thin. The free-swimming fry returns to the female's mouth at dust and when danger threatens, but gradually this instinct disappears and the young becomes fully independent.

(S) Red Tailed Black Shark

Scientific Name   :   *Labio bicolor*

Identification

Colour of the fish is black with orange caudal fin, caudal fin is bicurcated. Fish is having two pairs of barbs.

Full Length

12 cm

Figure 7.S.1: *Labio bicolor*

Food

Live as well as artificial foods. Augmented by plant foods.

Sex Ratio

One male : one female

Breeding

Breeding in captivity can be successful only after injecting the brood fish with carp hypophysis (2 mg Hypophysis per 100 g live weight)

Water Quality

Temperature	:	20–24°C
pH	:	6.5–7.0

Type of Breeding

Female deposit 1000 eggs of non adhesive type. Incubation period is observed 24 hours/22° C. After spawning the eggs float freely in the water.

Aquarium Behaviour

Aggressive and territorial fish and hence keeping mono-species tank.

Chapter 8
Marine Ornamental Fishes

Introduction

The aquarium trade is one of the most lucrative among all fishery related activities. Freshwater aquaria are in separate able components of most household in Europe, USA and Asian countries like Japan and China, Marine Ornamental Fish trade has already taken deep roots in international markets. Over 1500 species of aquatic animals and plants (80 per cent of which can be farmed or cultured) are sold world wide for aquarium purpose. It was estimated that Marine species make up only 9 per cent of the volume of the aquarium trade. However, due to their value, they represent 20 per cent of the revenue.

Indian states like Tamil Nadu, Maharashtra, West Bengal and Kerala are main centers for such business. A preliminary survey was done to identify different species of ornamental fishes and invertebrate available along the cost of Saurashtra (Gujarat) to provide base information (Jose and Fofandi, 2003). Amita Saxena (2003) has given the list of 45 marine aquarium fishes of India.(Table 8.2) Nick Dakin (1992) has given the list of marine fishes which can be spawned in captivity. (Table 8.3)

Marine Aquaria

Marine Aquaria are being used more and more in laboratories, particularly for research on physiological problems. As a general rule they are more difficult to maintain then freshwater aquaria. The most important points to be watched are condition and temperature of the sea water and type of materials used in the construction of tanks, pumps and ancillary apparatus.

Tank

Generally traditional tanks are prepared from rectangular iron frame with glass side embedded with bitumen. However, silicon-sealed glass aquaria is good for seawater usage. All plastic tanks have been popular in recent years because they are non-toxic, light and require no maintenance. Small plastic tanks (Up to 60 lit) manufactured from a single mould, are useful for handling fry and for treatment purposes. A problem with plastic tanks is that the surface can be easily scratched making the side opaque, so abrasive cleaning materials must be avoided.

The ideal tank is probably the all glass silicon sealed unit. The thickness of glass varies with the capacity of tank, which is as under.

Capacity	Thickness of Glass Preferred
Up to 100 lit	6 mm to 8 mm
100–200 lit	8 mm
200–1000 lit	12 mm
Above 1000 lit	24 mm

Tubes

Plastic tubing is cheap and readily available in a variety of size. Hard polyvinyl chloride and polythene have proved vary successful. Taps and valves should also be of polythene, but these should be inspected to ensure that they contain non metal parts, which might come in to contact with circulating water.

Seawater

If direct seawater is to be used, it should be used with proper filtration. Newly acquired natural seawater should be stored in dark for 2 to 3 weeks. This allows plankton organisms to die and drop to the bottom of the reservoir and give time to slit to settle.

Synthetic seawater also be used successfully and is often indispensable when the laboratory is a long distance form the sea. Several formula have been published but one of the most successful is shown in Table 8.1, which has been used in the University of Illinoism USA for maintaining marine invertebrates.

Table 8.1: Preparation of Synthetic Seawater

Salt	Gm/lit.
Sodium Chloride	27.2
Magnesium chloride	3.8
Magnesium sulphate	1.6
Calcium sulphate	1.3
Potassium sulphate	0.9
Calcium carbonate	0.1
Magnesium bromide	0.1

Source: Prakash and Arora, 1997.

The density of aquarium seawater should be kept at approximately 1.25 and this should be checked once a week.

Filtration-Sieving

In sieving a proportion of the relatively large particles in the water can be removed by trapping them with glass wool or nylon nappy, or small scale with filter paper. Nylon is preferable to glass wool, which some times fragment so that particles of glass enter the circulating water and because lodged in the gills of fishes. Sieving is of only minor importance in aquarium work.

Light

Aquarium can be light with natural or artificial light. With natural light the tanks tends to become coated with algal growth and this is not desirable unless the occupants include browsing herbivores such as limpets or certain sea-urchins. Tungsten lamps also tend to encourage algae and produce a lot of heat.

Fluorescent tubes are probably the most satisfactory from of lighting. They produce no perceptible heat, hence have a high lumen out put and are available in wide range of tones. For most purposes tubes, which are most suitable, preferably with an increase in the amount of red light.

Tropical coral fishes need bright light and so do sea–anemones, such as *Anemonia*, which have symbiotic algae.

So far as possible light fittings should be protected from corrosion by two to three coats of good resin. For small tanks the fittings can be hung over a glass to transparent plastics sheet which is allowed to rest with an air gap, on the rim of the tank. Lighting from side is recommended, except as a temporary measure when observation have to be made.

Cleaning of Aquarium

It is good practice periodically, say once every three months, to give tanks through disinfect ion, which helps to keep the fish free from disease. To do this, the fish must be removed and accommodated temporarily elsewhere. If the temporary accommodation is satisfactory the tanks may be filled with a solution of Potassium permanganate or Mercurochrome and left to stand overnight. If the tank cannot be spared for so long the sides bottom and particularly the corners should be thoroughly scrubbed with a strong solution one these substances, whilst the fitting, overflow or suction pipes, filter pumps etc should be steeped in it. It is most essential that the disinfectant should be well and thoroughly washed out before the fish are replaced, for all disinfectants in quite low concentrations are toxic. Washing out is tedious and often tanks longer than the disinfect ion but it must be done thoroughly or three is grave danger of losing the stock of fish.

The fishes, which are used for small and big aquariums are mentioned below.

Table 8.2: List of Marine Aquarium Fishes

Sl.No.	Scientific Name	Common Name
Fin Fishes		
1.	*Heterodontus francisci*	Californian horn shark
2.	*Triakis semifasciata*	Leopard shark
3.	*Gymnomuraena zebra*	Zebra moray
4.	*Gymnothorax meleagris*	Wide spot moray eel
5.	*Muraena pardalis*	Dragon face moray
6.	*Rhinimuraaena ambinensis*	Blue ribbon eel.
7.	*Teaenicongaer digneti*	Cortez garden eel
8.	*Plotosus anguillaris*	Striped catfish eel
9.	*Monocentris japonicus*	Pinecone fish
10.	*Myripristis murdjan*	Big eye squirrel fish
11.	*Adiryx suborbifalis*	Tris eel
12.	*Aeliscus strigatus*	Striped shrimp fish
13.	*Syngnathus leptorhynchus*	Bay pipe fish
14.	*Doryrhamphus melanopleures*	Fantail fish
15.	*Hippocampus kuda*	Golden sea horse
16.	*H. hudsonius*	Florida sea horse
17.	*Scorpeana guttata*	California scorpion fish
18.	*Pterois volitans*	Lion fish
19.	*Synanceia verrucosa*	Reef stone fish
20.	*Liopoproma rubre*	Swissguard basslet
21.	*Grammistes sexlineatus*	Golden striped grouper
22.	*Pseudochromis flanivertex*	Orange striped dottyback
23.	*Haemulon sexfasciatum*	Grayback grunt
24.	*Selene vomer*	Look down
25.	*Equetus lanceolatus*	Jakknife fish
26.	*Gramma loreto*	Royal gramma
27.	*Platax viola*	Rock croaker
28.	*Chaetodon faleiferi*	Scythe butterfly fish
29.	*Chelmon rostratus*	Copper banded angel
30.	*Holacanthus calrionensis*	Clarion angel fish
31.	*H. ciliaris*	Quin angel fish
32.	*Pygoplitis diacanthus*	Regal angel fish
33.	*Lubrisomus xanti*	Large mouth blenny
34.	*Exallias brevis*	Short bodies blenny
35.	*Lythrypneus dalli*	Blue handed goby

Contd...

Table 8.2–Contd...

Sl.No.	Scientific Name	Common Name
36.	Gobiosoma digueti	Banded clear goby
37.	G. puncticulatus	Read head goby
38.	G. multifasciatum	Green band goby
39.	Zanclus cornutus	Moorish idol
40.	Lo vulpinus	Fox face
41.	Balistapus undulatus	Undulated trigger fish
42.	Diodan holecanthus	Porcupine fish
43.	D. hystrix	Baltoon fish
44.	Bothus mancus	Pacific peacock floundes
45.	Pleuronichthys coenosus	C-oturbot.

Source: Amita Saxena, 2003.

Table 8.3: Marine Fishes which can be Spawned in Captivity

Egg-scatters	
Angel fishes	*Centropyge* spp.
	Holacantins spp.
	Pomocanthus spp.
Blennis	*Aspidontus* spp.
	Petroscirtes spp.
Butterfly fishes	*Cbaetodon* spp.
Manarin fishes	*Syncbiropus* spp.
Wrasses	*Pseudocheilinus* spp.
	Thalassoma spp.
Egg-depositors	
Anemon fishes	*Amphiprion* spp.
Damselfishes	*Abudefduf* spp.
	Dascyllus spp.
Gobies	*Gobiosoma*
	Oceanops
	Lytbrypnus dalli.
Grammus	*Gramma loreto*
Hawk fishes	*Oxycirrbiates* spp.
Mouth brooders	
Cardinal fishes	*Sphearannia* spp.
Jaw fishes	*Opistbognabus* spp.
Pouch brooders	
Seahorses	*Hippocampus* spp.

Source: Nick Dakin, 1992.

Marine Aquarium Keeping

Following are some of the important points to be kept in mind for sating up of marine aquarium:

It is important to note that before sating up salt water aquarium one should think about the reactions of salt water with metals.

1. No metal should come in contact with salt water.
2. No marine water aquarium should be smaller than 20 gallons capacity.
3. Collect only offshore sea water, inshore water is often polluted or heavily leaden with silt and bacteria.
4. It may also be too dilute after periods of heavy rainfall.
5. Do not set up the aquarium in front of window in direct sunlight or heavy growth of algae will occur.
6. Sea water weight is about 8.5 lb/gal. Therefore a 20 gallon tank contains at least 170 lb of water when full. In marine aquarium maximum carrying capacity is three inches of animal per square foot of filter bed surface area.
7. pH of marine aquarium water should be maintained at 7.5 to 8.3.
8. Temperature or marine aquarium should be kept between 21–23°C and then keep constant Marine fishes can't tolerate much fluctuation as tropical fishes can.
9. Regarding plants except sea water plants (seaweed) no plants are common to marine aquarium. So no plants should be important and introduced in to the aquarium.
10. Complete coverage by glass cover is most advisable for protecting against evaporation.
11. Acceptable ranges

    Unionized Ammonia: 0.01 ppm (0.1 ppm is usually lethal to the more delicate fishes and invertebrates)

    Nitrate: Less than 0.1 ppm as nitrate ion

    Nitrite: Less that 20.0 ppm as nitrite ion

    Dissolved oxygen: Not less than 1.0 ppm
12. Removal of uneaten particles every day is important.
13. Marine tropical fishes needs live food. Most marine fishes eat small quantity of *Tubifex, White worms*, Daphnia and bits or fish and clams. Brine shrimps are excellent food for them.
14. When they are treated for diseases use fresh water and permanganate instead of salt treatment.
15. Remove the fish as soon as any discomfort is observed.

Details of Marine Ornamental Fishes

Canion Pyjane or Cardinal Fish

Classification

Order	:	Perciformes
Family	:	Acanthuridae
Scientific Name	:	*Sphaerrua neratopterus*

Identification

The eyes are red and body is quite deep in relation to the length. The large head section, back to the first of the two dorsal fins, is yellow-brown in colour. A dark brown vertical band joins the first dorsal fin to the pelvic fins. A spotted paler area covers the rear of the fish. The large eye indicates a naturally nocturnal behaviour. The adult fish reaches to 3 inches size. They are hardy in nature and can easily be kept in marine aquarium. They are slow moving fish.

Length

100 mm in wild. However, rarely seen above 75 mm in aquarium.

Aquarium Behaviour

Do not keep with larger boisterous species.

Food and Feeding

They prefer fresh protein such as shrimp flesh or beef heart. However, they accept dried fish food also.

Powder Blue Surgeon

Family	:	Acanthuridae
Scientific Name	:	*Acanthus leacosternon*

Distribution

Indo-Pacific Ocean.

Identification

This is a favorite surgeon among aquarists. The oval shaped body is a delicate blue, the black of the head is separated from the body by a while area beneath the jaw line. The dorsal fin is bright yellow, as is the caudal peduncle. The white-edged black caudal fin carries a vertical white crescent. The female is larger than the male. In common with all surgeon, it requires plenty of space and optimum water condition.

Length

250 mm in wild and 180–200 mm in aquarium.

Figure 8.1: *Acanthus leacosternon*

Food and Feeding

Protein foods and vegetable matter, Bold grazer.

Aquarium Behaviour

Keep only one in the aquarium. Dealers usually segregate juveniles to prevent quarrels.

Golden Tang; Power Brown

Family	:	Acanthuridae
Scientific Name	:	*Acanthurus glaucoparies*

Distribution

Mainly the Pacific Ocean, but is some times found in the eastern Indian Ocean.

Identification

It is fairly easy to identify this fish by the white area on the checks. Yellow zones along the base of blue-edged dorsal and anal fins may extend in to the base of the caudal fin. A yellow vertical bar crosses the caudal fin.

Length

200 mm in wild.

Food and Feeding

Algae.

Figure 8.2: *Acanthurus glaucopairies*

Figure 8.3: *Acanthurus sohal*

Clown Surgeon Fish; Blue-lined Surgeon Fish

Family	:	Acanthuridae
Scientific Name	:	*Acanthurus lineatus*

Distribution

Indo-Pacific Ocean.

Identification

The yellow ground colour of the body is covered with longitudinal dark-edged

Figure 8.4: Marine Clown Fish

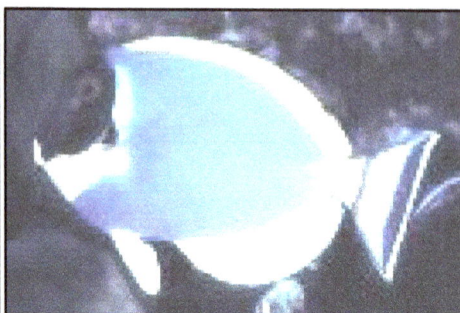

Figure 8.5: *Acanthurus lineatus*

light blue lines. The pelvic fins are yellow. This fish has a split level of colouration; this is a lighter area to the bower body with decorative parallel longitudinal lines above like other surgeon fishes, it appreciates some coral or rock work to provide welcome sheltering places.

Length

280 mm in wild.

Food and Feeding

Algae.

Aquarium Behaviour

Keep only one of these fish in an aquarium.

Bennetts Sharp Nose Puffer

Classification

Order	:	Tetrodomliformes
Family	:	Canthigasteridae
Scientific Name	:	*Canthigaster bennetti*

Identification

This species is most colourful and suitable for marine aquarium and very pleasant to look. It is a hardy fish. Size of this fish is 2 inches. It is found abundantly throughout the Indo-Pacific. It is peaceful fish.

Food

It feeds on fleshy fish. However, it adopts new condition and food offered.

Sunburst Butter Fly Fish

Order	:	Perciformes
Family	:	Chaetedontidae
Scientific Name	:	*Chaetodon klieni*

Figure 8.6: Flying Fish

Distribution

Indo–Pacific

Identification

A black bar runs down from the tip of head down each side and through the eyes, as it does in all member of this genus. The mouth and head is white in colour. Body golden in colour with silver spots. It is quite hardy and relatively easy to keep in aquarium.

Food

Food is no problem and all normal aquarium foods are suitable.

Blue Striped Butterfly Fish

| *Family* | : | Chaetodontidae |
| *Scientific Name* | : | *Chaetodon frembii* |

Identification

The yellow body is marked with upward slanting diagonal blue lines. A black mark appears immediately in front of the dorsal fin and the black of the caudal peduncle extends in to the rear of the dorsal and anal fins. The caudal fin has white, black and yellow vertical bars. This butterfly fish lacks the usual black bar through the eye.

Figure 8.7: *Chaetodon frembii*

Length

200 mm in wild.

Aquarium Behaviour

Clam community fish. Aquarium breeding is not observed.

Food and Feeding

They are grazing fish. They feed on algae, sponges and corals. It is necessary to feed several times in a day in aquarium. They also take zooplanktonic food.

Breeding

There is no external difference between the sexes, although at breeding times the female may become noticeably swollen with eggs. The fertilized eggs float briefly until they hatch. The larvae then feed and develop in the planktonic layers for several months before migrating back down the reef floor.

Long Nosed Hawk fish

Family	:	Cirrhitidae
Scientific Name	:	*Oxycirrhites types*

Distribution

Indian Ocean mainly.

Identification

The elongated body of this hardy fish is covered with squared pattern of bright

Figure 8.8: *Oxycirrhites types*

red lines. The snout is very long and suited to probing the coral crevices for food. There are small cirri or frostlike growth at the end of each dorsal fin spine and on the nostrils. The female is larger than the male and the male has darker red lower jaws. There are black edges to the pelvic and caudal fins. They rest on a piece of coral and wait for foods to pass by.

Length

100 mm in wild.

Food and Feeding

Most frozen marine foods, sits on corals or a rock, the dashes out to grab food.

Aquarium Behaviour

Peaceful can be kept in small groups. They will appreciate plenty of "Perching Places" in the aquarium.

Breeding

In nature, spawning occurs from dusk on wards. Reports of aquarium spawning suggest that the female lays patches of adhesive eggs after courtship activity. The eggs are laid and fertilized on a firm surface, where they subsequently hatch.

Neon Goby

Family : Gobiidae

Scientific Name : *Gobiosoma oceanops*

Figure 8.9: *Gobiosoma oceanops*

Identification

Two characteristic distinguish this most familiar goby; the electric blue colouration of the longitudinal line on the body and the cleaning services it offers to other fishes. The species can be positively identified by the gap visible between the two blue lines on the snout when the fish is seen from above.

Length

60 mm in wild and 25 mm in aquarium.

Food and Feeding

Parasites, small crustaceans and plankton, bottom feeder.

Breeding

G. oceanops has been bred in the aquarium and is now regularly bred on a commercial basis. Before spawning the male's colour darkens and he courts the female with exaggerated swimming motions. Assuming a position on the aquarium floor until the female takes notice of him. Spawning activity occurs in a cave or other similar sheltered area. The fertilized eggs hatch after 7–12 days. In the wild, the fry fed on planktonic foods for the first few weeks. In the aquarium, start the fry off with cultured rotifers followed by newly hatched *artemia nauplii*. Life span of this species is only a year or two.

Yellow Goby

Family : Gobiidae

Scientific Name : *Gobiodon okinawae*

Distribution

Pacific Ocean.

Figure 8.10: *Gobiodon okinawae*

Identification

Yellow Goby is small in size but highly territorial. They are very poor swimmer and prefer to remain perched on favourite vantage points. Once settled, it shows little concern for other inhabitant that poses no threat. Their bodies are elongated. The head blunt with high set eyes. Poor water quality is not tolerated and the fish will refuse to eat and may turn dirty brown colour as a result. If no improvement is made death usually follows.

Length

30 mm in wild.

Aquarium Behaviour

Very peaceful, except with the same species.

Food and Feeding

They will readily accept small particles of most marine fare, including frozen, live and flake foods. Live foods are particularly relished, especially when the fish is first settling in aquarium.

Breeding

They lay eggs and protected by males.

Lemon Goby

Family	:	Gobiidae
Scientific Name	:	*Goblodon citrinus*

Figure 8.11: *Goblodon citrinus*

Distribution

Indo-pacific.

Identification

The colours of this fish are quite brilliant. Their bodies are elongate, the head blunt with high set eyes. Sexing gobies can be different, although females may become distended with eggs at breeding time and there are the typical difference in the size and shape of the genital papillae.

Length

30 mm in wild.

Food and Feeding

Once settled, will accept most marine foods of a suitable size, particularly fond of live foods.

Breeding

Spawning occurs in burrows or in sheltered areas, with eggs being guarded by the male.

Royal Gramma

Family	:	Grammidae
Scientific Name	:	*Gramma loreto*

Identification

The main feature of this species is its remarkable colouring. The front half of the body is magenta, the rear half bright golden-yellow. A thin black line slants back wards through the eye.

Length

130 mm in wild, however in aquarium it grows up to 75 mm only.

Food and Feeding

They eats a wide range of foods, including chopped shrimp and live brine shrimp. They also accept flake foods.

Breeding

Spawning activity has been observed in captivity. Four grouped themselves in to two "Pairs", each comprising one small and one large fish. The larger fish lined a pit in the sand with stands of large glued together with glandular secretion and pair spawned "stickleback" fashion. The smaller fish was enticed into the pit several times, closely followed by the larger of the two, who then stood guard over the pit fill of eggs.

In this species males are larger than females of the same age group.

Black Cap Gramma

Family	:	Grammidae
Scientific Name	:	*Gramma melacare*

Identification

The colour of body of this fish is purple. They have black patch over the crown of the head.

Length

100 mm in wild.

Food and Feeding

They prefer live foods but will usually accept frozen marine fare after and initial setting in period. This species need a little more care and may need to be tempted with live brine shrimp and Mysis initially.

Aquarium Behaviour

This most attractive is extremely territorial and should be kept in the absence of its own or similar species. They are shy and secretive by nature. Requires a comprehensive arrangement of rockwork in which to hide.

Wrasses

Family	:	Labridae
Scientific Name	:	*Cirrhilabrus rubriventralis*

Distribution

Indian Ocean

Identification

This is an extremely attractive species with distinct male and female colouration. The male is much more intensely coloured than the female and has an elongated leading edge to the dorsal fin. In suitably sized tanks its is possible to keep a pair, or one male and several females, together. In this arrangement, the male will intensify his colours and display to the females by dashing in front of them flaring his fins. Optimum water conditions are essential for this fish. They are peaceful community fish.

Food and Feeding

They will accept most marine fare, including frozen, live and flake fish.

Genital Papillae

They are breeding tubes that extend from the vent of each fish; usually larger in females than males.

Pyjama Wrasse

Family	:	Labridae
Scientific Name	:	*Pseudocheilinus hexataenia*

Figure 8.12: *Pseudocheilinus hexataenia*

Identification

P. hexataenia has many of the attributes of an ideal aquarium fish; it rarely bothers invertebrates or other fish (as long as they care not of its own kind, or similar), it is colourful, easily fed, interesting in its activities and generally very inexpensive. This disease resistant species can be heartily recommended to any new comer to the hobby.

Aquarium Behaviour

They are peaceful, except with the same or similar species.

Length

50 mm in wiid and 70 mm in aquarium.

Food and Feeding

They readily accept most marine frozen, live and flake foods.

Yellow Headed Jaw Fishes

Family	:	Opistognthidae
Scientific Name	:	*Opistognathus auriform*

Identification

The delicately coloured yellow head is normally all you see of this fish, but the rest of the body is an equally beautiful pale yellow. The eyes are large. It needs a reasonably soft substrate in which to excavate a burrow, entering the hole tail first at any sign of trouble. Like other jaw fish, it is good jumper.

Length

125 mm in wild.

Aquarium Behaviour

Peaceful and rarely disturbed by other fishes.

Food and Feeding

Finely chopped shellfish meat. Makes rapid lunges from a vertical hovering position near its burrow to grab any passing food.

Figure 8.13: *Opistognathus auriform*

Two-banded Anemonefish

Family	:	Pomacentridae
Scientific Name	:	*Amphirion clerkii*

Distribution

Indo-Pacific.

Identification

The species is highly variable in colouration depending on its location, but generally the body is predominantly dark brown but for the ventral regions, which are yellow. All the fins, with the exception of the paler caudal fins, are bright yellow. Two tapering white vertical bars divide the body in to thirds; the juvenile forms, there is a third white bar across the rear of the body.

Figure 8.14: *Amphirion clerkii*

Length

120 mm in wild; 75 mm in aquarium.

Aquarium Behaviour

It is peaceful and makes an ideal choice for a mixed aquarium. They require excellent water quality.

Food and Feeding

They are constant feeder and requires food for 3 to 4 times a day. They eat small crustaceans, small live foods, algae, vegetable based foods, bold feeder.

Green Chromis

Family	:	Pomacentridae
Scientific Name	:	*Chromis caerulea*

Distribution

Indo-Pacific Ocean.

Identification

This hardy colourful shoaling species has a brilliant green blue shine to the scales. The caudal fin is more deeply forked than in some damsel fishes. Keep these

Figure 8.15: *Chromis caerulea*

fishes in shoals; individuals may go into decline. Ideally, a shoal should consist of at least six fishes. It is lively and attractive species and can safely be kept in mixed fish and invertebrate set–up.

Length

100 mm in wild and 50 mm in aquarium.

Aquarium Behaviours

Generally peaceful.

Food and Feeding

Chopped meat.

Breeding

Normally, there are no clear determining characters between the sexes. There is one method of determining sex by external observation–a technique similar to that used for determining sex in freshwater Cichlids and that is by looking at the genital papillae (often called the ovipasitor). The male genital papilla is narrower and more pointed than the female's. However, it can best be seen during breeding activity is noticed, when the papillae are easier to see. Spawning in damselfish entails the selection guarding of eggs.

Angel Fish, Bicolor Cherub, Oriode Angel

Family   :   Pomacanthidae

Scientific Name   :   *Centropyge bicolor*

Identification

The rear point of the, from behind the head as far as the caudal fin, is bright purple-blue. The small bar across the head over the eye is the same bright shade,

Figure 8.16: *Centropyge bicolor*

while the head and caudal fin are yellow. In groups a solitary, male will dominate a "harem" of females. If the male is removed from the group or dies, then one of the females will change sex to replace him. This procedure occurs every time the group becomes "male-less". This fish is susceptible to disease. Use copper remedies with care.

Length

125 mm in wild.

Food and Feeding

Meat foods and plenty of green stuff.

Aquarium Behaviour

They are peaceful, provided plenty of hiding places are available.

Flame Angelfish

Family	:	Pomacanthidae
Scientific Name	:	*Centropyge loriculus*

Distribution

Pacific

Identification

The fiery red-orange body has a central yellow are crossed by vertical dark bars. The dorsal and anal fins are similarly dark–tripped. The flame Angelfish is not difficult to keep but does require excellent water conditions and varied diet. Although generally expensive, this is a most rewarding fish to keep.

Figure 8.17: *Centropyge loriculus*

Length

100 mm in wild.

Aquarium Behaviours

This is a peaceful fish.

Food and Feeding

Meat foods and Plenty of green stuff grazer.

Blue Ring Angelfish

Family	:	Pomocanthidae
Scientific Name	:	*Pomacanthus annularis*

Distribution

Indo–Pacific Ocean.

Identification

Blue line runs from either side of the eyes diagonally across the brown body. The lines rejoin at the top of the rear portion of the body. A dominant blue ring lines behind the gill cover. Juveniles are blue with a distinctive pattern of almost straight transverse white lines. Juveniles and adults were once considered to belong to different species.

Length

400 mm in wild; 250 mm in aquarium.

Aquarium Behaviour

Suitable when very young, but become destructive with age.

Food and Feeding

Meat foods and green stuff grazer can be given as food.

Figure 8.18: *Pomacanthus annularis*

Figure 8.19: Marine Angel (*P. semicirculatus***)**

Sea Horse

Family	:	Syngnathidae
Scientific Name	:	*Hippocanpus erectus*

Identification

The pelvic and caudal fins are absent and the anal fin is very small. The tail is prehensile. The colouration of this species is variable; individuals may be gray, brown, yellow or red. The male incubates the young in the abdominal pouch. This species is also frequently referred to as *H. budsonitus*.

A pale individual of this elegant species, seahorses adopt a vertical position when at rest. When swimming they lean forward, propulsion being provided by the fanlike dorsal fin.

Length

150 mm in wild.

Aquarium Behaviour

Needs quite, non-boisterous companions.

Food and Feeding

Small animal food.

Figure 8.20: *Hippocanpus erectus*

Figure 8.21: *Hippocanpus huda*

Breeding

When seahorse reproduce, the female uses her ovipositor tube to deposit the eggs in to the male's abdominal pouch, where they fertilized and subsequently incubated. Incubation period ranges from two weeks to two months.

Asian Dragon Fish

Family	:	Osteoglossidae
Scientific Name	:	Arowana

Identification

The fishes of this family are widely regarded as the "King of the aquarium" due to its popularity. It is also considered to be another "Feng-Shui" fish bringing luck wealth and prosperity. It is unique in shape with large mouth and neatly arranged distinctive scales. The body is symmetrical with limb–like fins.

Distribution

Indonesia, Vietnam, Philippines, Malaysia, Cambodia.

Protected as an endangered fish.

Because of the great demand due to body colour and shape, colour of scales and their pattern, fish shape, eyes, mouth, teeth, gill covers, vent and swimming pattern, they have been hunted in the wild, resulting stock depletion. Conservation efforts have been made and the Asian dragon is now protected under CITES Category–1 as an endangered fish. Trading of the fish is regulated by CITES, with breeders having to show that they have successfully bred the fish in captivity only. F-1 generations are allowed to be marketed. Throughout the world only a few farms located in Indonesia, Malaysia and Singapore have managed to get the CITES license.

Red Arowana (Red Dragon)

It is a most preferred variety due to its red colour. A good quality fish has a chilli or blood red body colour and the edges of the scales are radiant. Though the scales orange yellow with a tinge of light green in the young, it has prominent red trimmings

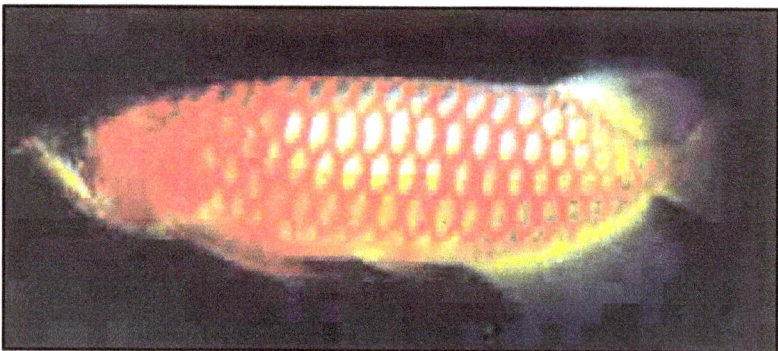

Figure 8.22: Red Arowana

as the fish grows. The scales have a shade of purple blue when it becomes fully grown.

Water quality, Food and feeding as well as breeding is same as Golden Arowana.

Golden Arowana

> *Family*   :   Osteoglossidae

The value of this fish depends on the body colour of the fish. More uncommon the colour, the more valuable is the fish. The golden dragon fish is the most expensive at present. There are more than six varieties, based on geographic distribution.

Figure 8.23: Golden Arowana

Native

The golden Arowana is a native of West Malaysia.

Identification

As the fish matures, the scales turn yellow with a tinge of olive green which extends to the fourth raw of scales from the stomach region. The edge of each scale is pinkish, with some golden yellow colour. As the fish is fully grown the pink colour of the scales diminishes and the whole fish turns golden yellow, extending to the fifth row of scales and over the back. However, depending on the condition of the environment in which they are grown, the colour may differ.

Water Quality

> *Temperature*   :   26–30° C
>
> *pH*   :   6.5–7

They prefer neutral to slightly acidic water.

Food and Feeding

In the wild they eat a wide range of foods.

Breeding

They are difficult to breed in aquarium conditions and breed only in shaded natural earthen ponds planted with aquatic plants.

Spawning takes place almost throughout the year with the peak season in July and December. Sexes are indistinguishable before maturity and sex differentiation is difficult even after maturity. Males are slightly slimmer than females and possess deeper and wider mouths. A brooding male can be recognized easily by its brood pouch used for holding eggs. Due to the aggressive nature of the fish, the growth of juveniles is often uneven; hence it is better to isolate the individual fishes.

Red-Tailed Gold Dragon

It is native of Indonesia, is more affordable than the Golden dragon. This is because the dorsal portion of the fish dark green including the dorsal fin and upper half of its tails fin with the rest of the body scales gold. In younger fish the scales are golden tinged with pink; however, the luster of the scales will not extend to the fifth row as in the case of the Gold Dragon.

Green Dragon

The green Dragon is a native of Thailand, Malaysia, Vietnam and Myanmar. It has green scales and a very distinct lateral line. It is the cheapest among the Dragon fish. Due to its widespread distribution there could be many differences in its appearance and colour pattern. The more expensive varieties have purplish–spotted scales. Fish without purple spots are very common and regarded as a cheap variety.

Striped Cat Fish

Scientific Name	:	*Mystus vittatus*
Synonyms	:	1. *Silurus nittatus*
		2. *Macrrones tengare*
		3. *Mystus tengare*
		4. *M. aartifasciatus*

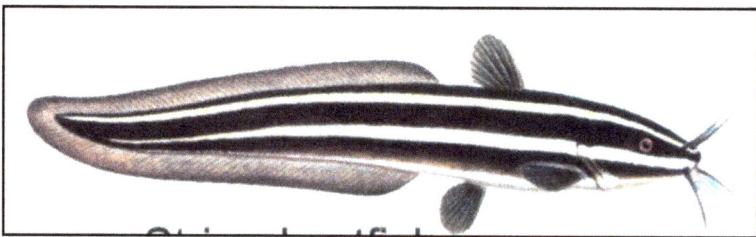

Figure 8.24: Striped Catfish

Distribution

India, Pakistan and Burma.

Identification

The body has no scales. The upper and lower jaw and the palatal bone are equipped with teeth. There are two pairs of barbells on the upper and lower lip. The

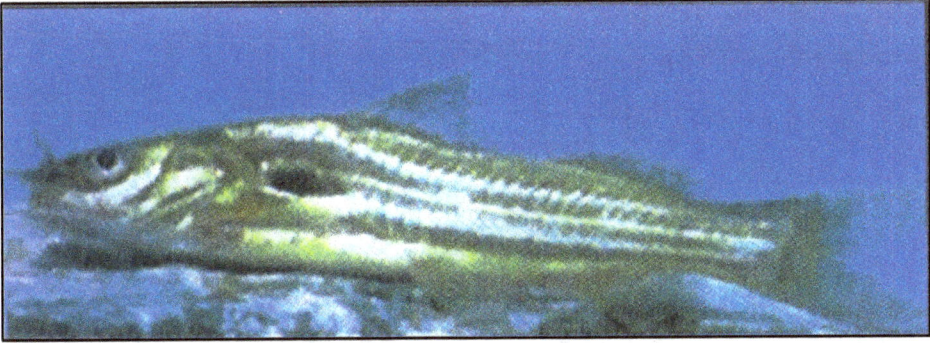

Figure 8.25: *Mystus vittatus*

dorsal fin inserts relatively near the front end, the first ray of the dorsal fin is transformed in to a strong spine that is saw edge. The adipose fin is smaller and extended lengthwise; the anal fin is small. The eyes are also small and often covered with skin.

Full Length

20 cm.

Food

Live as well as artificial foods including live fish.

They can be kept in community tank, with fish bigger than 8 cm.

Breeding

Brood fish should be kept only with sufficient hiding and dim lighting. They should be kept in mono species tank. Sexual dimorphism for mare and female is not known. They are egg layer fish. Breeding in captivity is not known. However, spawning in the wild takes place amidst the tangle of roots and aquatic plants on the bottom, where the female deposits eggs.

Chapter 9

Diseases and their Treatment of Ornamental Fishes

Introduction

Most of the diseases that affect live bearers are fairly easy to recognize. Most of them are usually parasitic or bacterial and signs are found on the external surface. There are some internal diseases also and very difficult to treat. Fish diseases can be predicted or identified either by fish swimming movement or from skin problems. Some time white spots are seen, sometime it is slimy; sometime some parasites are seen on skin or on gill. This observation leads you to come to the conclusion of fish diseases. With the help of Chart–A, we can identify fish disease from swimming problems. Chart–B will guide you to identify disease because of skin problems.

Over all signs for disease affected fish are as under:

1. The sign of illness is loss of appetite but it can be easily missed in well populated tank.

2. Sick fish may hang around looking rather dull.

3. Movement if gills become fast. A fish symptom like this sometimes spends time near the surface of the water or close to the pump outlets. This is due to gill problems.

4. Very often fish looses its colour and becomes pale due to presence of excess mucus caused by parasite on the skin. In long standing cases of stress they may look darker.

5. The change of the shape of body also indicates the presence of disease in fishes; a hollow or dropsy can produce a swallow belly. Odd bumps may suggest presence of parasites in the tissues. Sometimes fin erosion also found.

6. A fish may flick against hard objects such as rocks, to scratch out external parasite.

7. Difficulty in balance maintenance or abnormal posture suggests a serious parasitic or bacterial infection. A slight undulating swimming on the spot is also found. This suggests chilling of water or change in nature of water. It is also found in live bearers affected by columnar.

Similarly other signs of illness are as under:

Signs of Illness

Before we look at some specific diseases and conditions that may affect aquarium fishes, here we review the general sings of illness that aquarium fishes may show:

Loss of Appetite

This is often the first sign of ill health, but in a well populated tank it is easy to miss.

Restlessness

Sick fishes may hang around looking rather dull. This also occurs fairly early in serious diseases and is easily missed because the fish may hide.

Rapid Gill Movements

A fish with this symptom will often spend time near the surface of the water or close to pump outlets. If usually signifies a gill problem. If there is no obvious skin damage, it may be *Oodinium* (Velvet disease) or a bacterial infection.

Change in Colour

Fish often become pale if actually distressed or they may look pale as the result of excess mucus caused by parasites on the skin. In long standing cases of stress they may become darker.

Change in Shape

All kinds of things may change the shape of a fish. Wasting and a hollow bellied appearance may suggest tuberculosis, or a female that has simply worked hard producing fry. A swollen belly can be caused by pregnancy or dropsy. Odd bumps may be due to the presence of parasites in the tissues; this is more common in wild caught specimens. Fin erosions also come under this heading, since they do cause a change in the fishes' appearance.

Flashing

A fish may flick against hard objects, such as rocks, as if scratching or trying to dislodge parasites. It suggest slime disease, skin flukes etc.

Abnormal Posture

Difficulty in maintaining balance or an abnormal posture suggests a serious parasitic or bacterial infection.

Shimmkying

This is a strange sideways undulating "swimming of the spot". It is usually suggests chilling or reaction to a change in water conditions. It is also seen in livebearers infected with columnar is (caused by *Flexibacter columnaris*) and is especially common in the larger sail fin Mollies.

(*Source*: Peter Scoff, 1987)

Common Skin and Gill Disease

(1) Freshwater White Spot (ICH) (up to 1 mm to 0.04 inch in diameter) (*Ichthyophthirius multifilis*) (Figure 9.4 and 9.5)

It is known as Ich, Ick, or white spot disease. It is the most common disease in home aquarium. White spot disease is caused by a protozoa called *Ichthyophthirius multifilis*. This is most familiar disease of all aquariums. The parasite is easily visible with necked eyes as white spot on fish. It is necessary to kill it before formation of cyst. Mature parasite develops large holes in the skin which gets infected by bacteria and fungi secondarily.

There are three phases to the life cycle of this protozoon. It is important to know its life cycle because white spot (ICH) is susceptible to treatment at only one stage of the life cycle.

(*i*) Adult Phase

It is embedded in the skin or gills of the fish, causing irritation (with the fish showing signs of irritation) and the appearance of small white nodules. As the parasite grows it feeds on red blood cells and skin cells. After a few days it bores itself out of the fish and falls to the bottom of the aquarium.

(*ii*) Free Swimming Phase

If a host is not found within 2 to 3 days, the parasite dies. Once a host is found the whole cycle begins a new. These three phases take about 4 weeks at 70° F but only 5 days at 80° F. For this reason it is recommended that the aquarium water be raised to about 80° for the duration of the treatment. If the fish can stand it, raise the temperature even higher up to 85°.

(*iii*) Cyst Phase

After falling to the bottom, the adult parasite forms into a cyst with rapid cell divisions occurring. After the cyst phase, about 1000 free swimming young comes out and looking for a host.

Symptoms

Salt-like specks are found on the body/fins. Excessive slime can be seen. Fish shows problems breathing (ICH invades the gills), clamped fins, loss of appetite and shows *Abnormal* behaviour such as unusual swimming patterns, refusing all food etc.

Treatment

The free swimming phase is the best time to treat with chemicals. Raising the aquarium temperature to 80° F will greatly shorten the time for the free swimming phase to occur. The drugs of choices are:

1. Quinine hydrochloride at 30 mg per liter.
2. Quinine sulphate can be used if the hydrochloride is not available.
3. Some aquarists like to use malachite green, but it tends to stain the plastic and silicone in the aquarium. Most commercial remedies contain malachite green and/or copper, which are both effective.
4. Use proprietary treatment and remove any activated carbon from filter.
5. Formalin (add small amount of malachite green, oxalate crystals etc to give it colour) 1:4000 for one hour or acetic acid 1:500 for 1 minute. Treatment should be given for 3–4 consecutive days.
6. Salt in 3 per cent solution is also effective.

(2) Gill Flukes (*Dactylogyrus*) (Figure 9.14 and 9.16)

Disease Flukes

External and internal "worm" parasites belonging to the classes Trematoda and Cestoda species in the family Dactylogyridae parasitise the gills and digestive tract. The family Gyrodactylidae parasitise affects the body and finnage. Cestodes parasitise affects the internal digestive tract. Rarely visible to necked eyes. They can be seen with good hand lens or microscope. They use their many-hooked foot for movement around gill membrane. Flukes are parasite affecting fish. They range from 0.05 to 3.00 mm long and there are actually a huge number of species in the genus.

Gyrodactylus–This is also one of the important Skin Fluke. Fish suffering from infestations of gill flukes may suffer respiratory problems as the flukes begin to damage the delicate gill tissues.

Secondary bacterial infection often occurs in fish left suffering from these parasites, due to the physical damage caused by the anchors.

Chemical control of both types of fluke can be achieved with Chloramine T, Malachite Green Formalin and Masoten, or Potassium Permanganate.

In order to kill all generations, repeat treatments may be necessary, the frequency being dependent on temperature and chemical.

Symptoms

In early stages (first 24/48 hours) fishes show signs of extreme skin/gill irritability, continually scratching and scraping on rocks etc., and "flicking" the pelvic and dorsal fins against the side of the body.

Treatment

Give short-term formalin baths (use 35 per cent solution of formaldehyde approximately) at a dilution of 0.2 ml per liter of water. This can be used for 1 hour if fish gets distressed stop treatment. Bath can kill adult flukes but repeated treatment is needed for egg removal. Individuals should be treated with hospital tank. Remember some eggs can logged in gills, which can fall out and cause development therefore complete cleaning of tank is important.

(3) Skin Flukes (*Gyrodactylus*)

Gyrodactylus

This is also one of the important Skin Fluke. These are found on skin Fish suffering from infestations of gill flukes may suffer respiratory problems as the flukes begin to damage the delicate gill tissues are livebearers. Each adult may contain several generations at different stages of maturity inside.

Secondary bacterial infection often occurs in fish left suffering from these parasites, due to the physical damage caused by the anchors.

Argulus (Figure 9.8)

This is also one of the important parasite found on skin. The Fish Louse it is easy to detect with the naked eye especially against the background of fins. Size varies from between 1mm and 5mm. Attaching themselves to the fish by suckers which damages the skin, they also inject a poison into the body of the fish which causes inflammation, bleeding and potentially secondary bacterial infection.

Lernaea–Anchor Worm (Figure 9.13)

Lernaea is a common parasite which is clearly visible to the naked eye and can reach 10 to 12mm. The parasite burrows its head into the fish tissue, under a scale and only the body and tail are normally visible. Lernaea lay eggs which can lay undetected in the pond and can hatch when conditions and water temperatures are right. Chemical treatments will not affect the viability of eggs so repeat treatments may be required to kill all generations.

Treatment

Use short-term formalin baths. This is sufficient. Chemical treatments recommended to eradicate these parasites are Masoten, Dimilin or Paradex. Chemical control of all types of fluke can be achieved with Chloramine T, Malachite Green Formalin and Masoten, or Potassium Permanganate. In order to kill all generations, repeat treatments may be necessary, the frequency being dependent on temperature and chemical. To disinfect, use acriflavine (trypaflavine) a 0.2 per cent solution at the rate of 1 ml per liter. Add antibiotic chloromycetin or tetracycline 250 mg capsule/ 50 l water.

(4) Slime Disease (*Chilodonella, Chclochaeta, Costia*) (Figures 9.6 and 9.10)

Disease caused by various external protozoan parasites (such as Ichtyobodo/ Costia and Chilodonella) and monogenetic flukes like Gyrodactylus.

Chart–A
To Identify Fish Disease from Swimming Problems

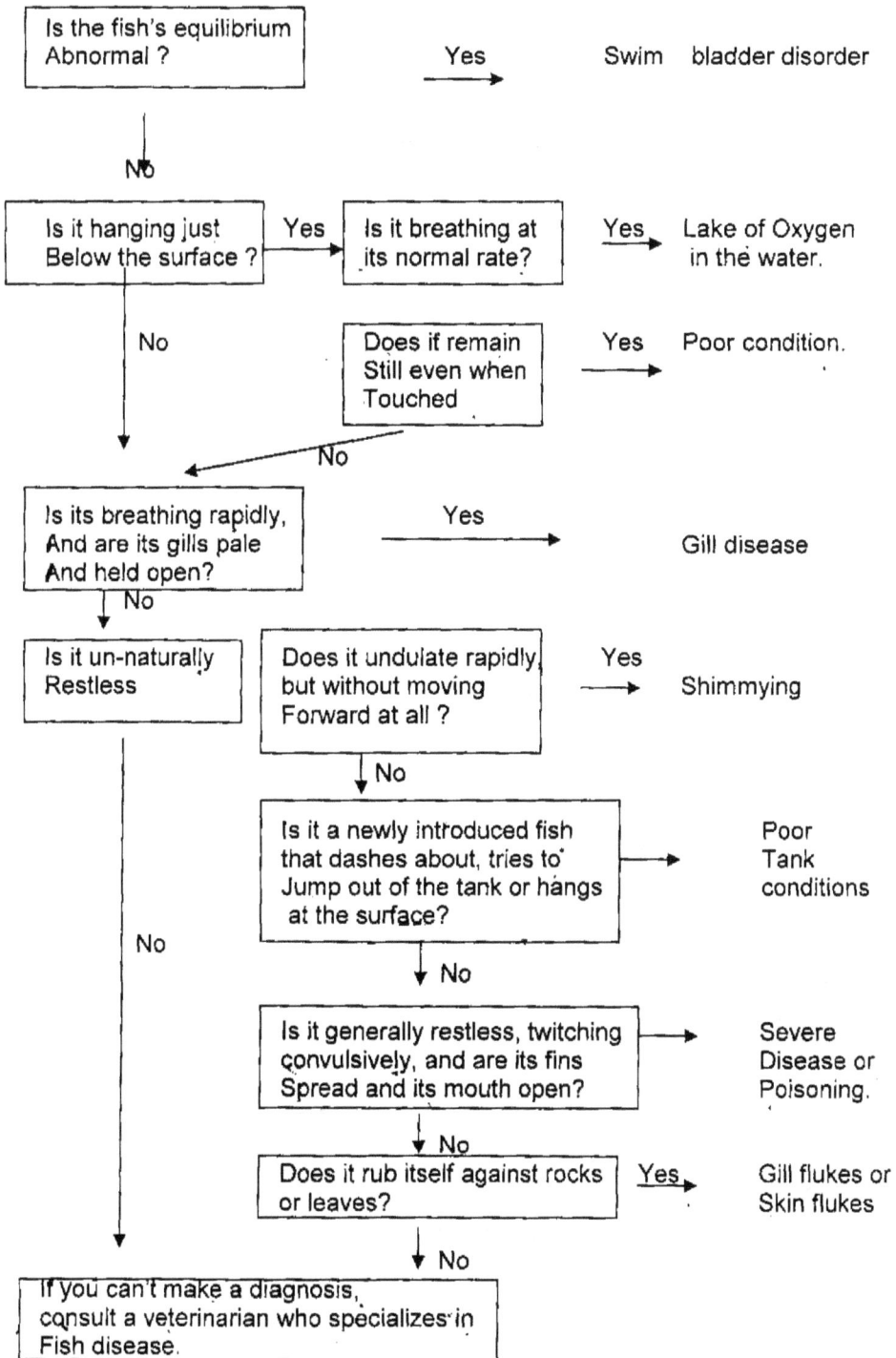

Is the fish's equilibrium Abnormal ? — Yes → Swim bladder disorder

No ↓

Is it hanging just Below the surface ? — Yes → Is it breathing at its normal rate? — Yes → Lake of Oxygen in the water.

No ↓

Does if remain Still even when Touched — Yes → Poor condition.

No

Is its breathing rapidly, And are its gills pale And held open? — Yes → Gill disease

No ↓

Is it un-naturally Restless

Does it undulate rapidly, but without moving Forward at all ? — Yes → Shimmying

No ↓

Is it a newly introduced fish that dashes about, tries to' Jump out of the tank or hangs at the surface? → Poor Tank conditions

No ↓

Is it generally restless, twitching convulsively, and are its fins Spread and its mouth open? → Severe Disease or Poisoning.

No ↓

Does it rub itself against rocks or leaves? — Yes → Gill flukes or Skin flukes

No ↓

If you can't make a diagnosis, consult a veterinarian who specializes in Fish disease.

Chart–B
To Identify Disease from Skin Problems

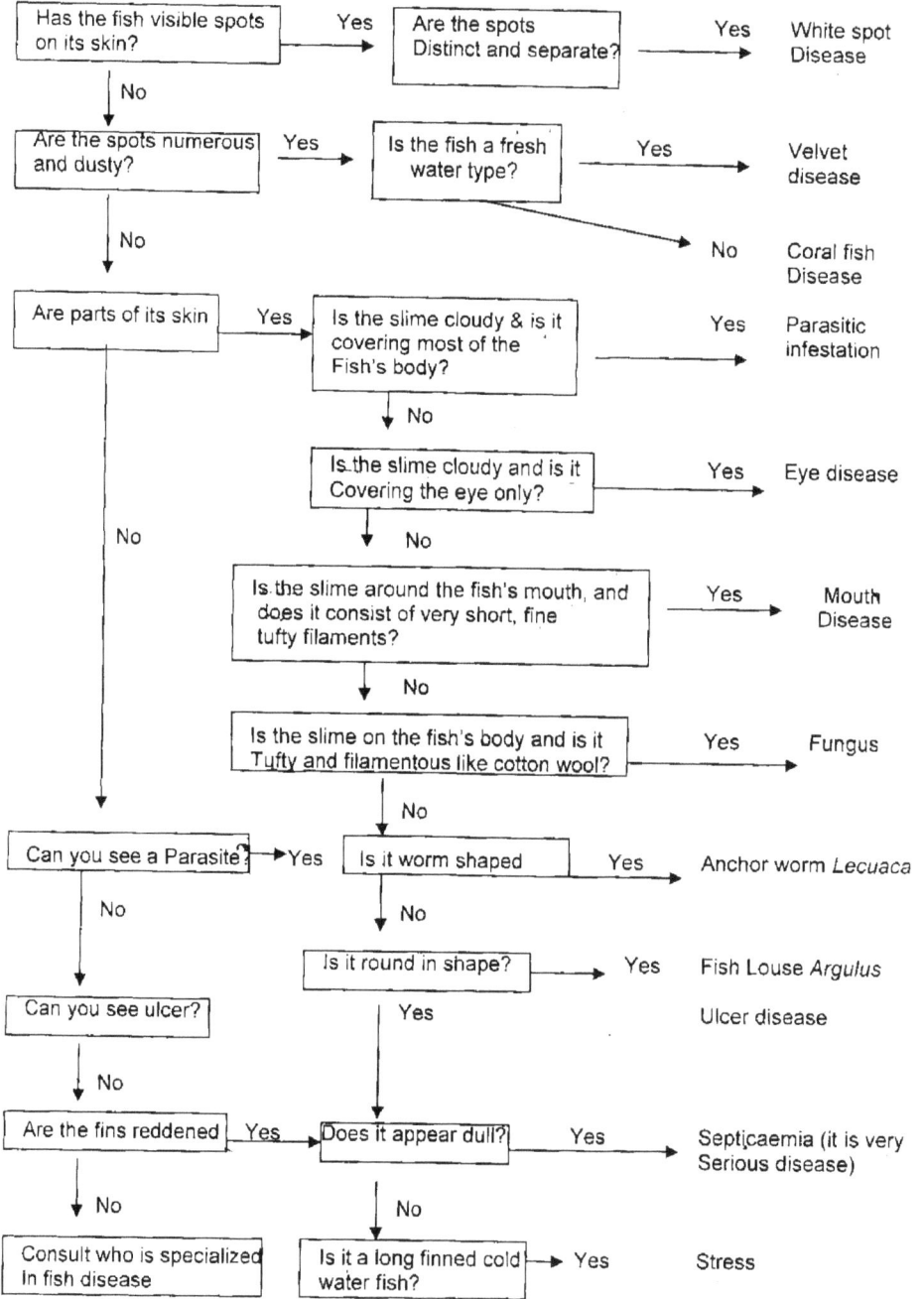

Has the fish visible spots on its skin?	→ Yes → Are the spots Distinct and separate?	→ Yes → White spot Disease

↓ No

Are the spots numerous and dusty? → Yes → Is the fish a fresh water type? → Yes → Velvet disease

→ No → Coral fish Disease

↓ No

Are parts of its skin → Yes → Is the slime cloudy & is it covering most of the Fish's body? → Yes → Parasitic infestation

↓ No

Is the slime cloudy and is it Covering the eye only? → Yes → Eye disease

↓ No (No, on left for "Are parts of its skin")

Is the slime around the fish's mouth, and does it consist of very short, fine tufty filaments? → Yes → Mouth Disease

↓ No

Is the slime on the fish's body and is it Tufty and filamentous like cotton wool? → Yes → Fungus

↓ No

Can you see a Parasite? → Yes → Is it worm shaped → Yes → Anchor worm *Lecuaca*

↓ No | ↓ No

Can you see ulcer? | Is it round in shape? → Yes → Fish Louse *Argulus*

↓ No | ↓ Yes → Ulcer disease

Are the fins reddened → Yes → Does it appear dull? → Yes → Septicaemia (it is very Serious disease)

↓ No | ↓ No

Consult who is specialized In fish disease | Is it a long finned cold water fish? → Yes → Stress

Different Types Fish Diseases

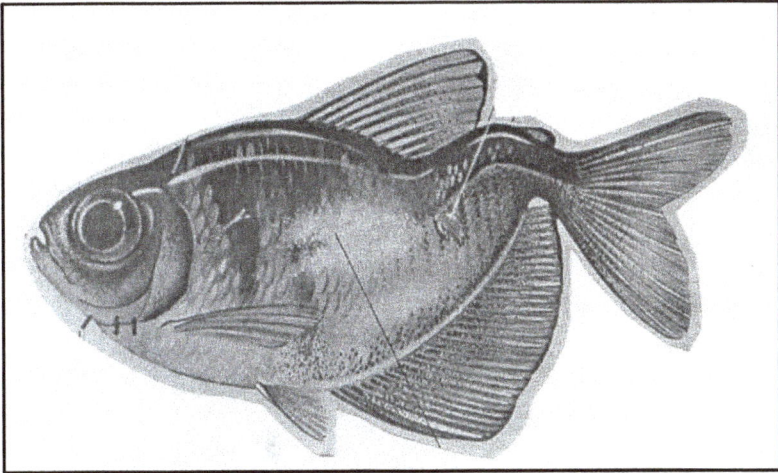

Figure 9.1: Bent Spine (Scoliosis)

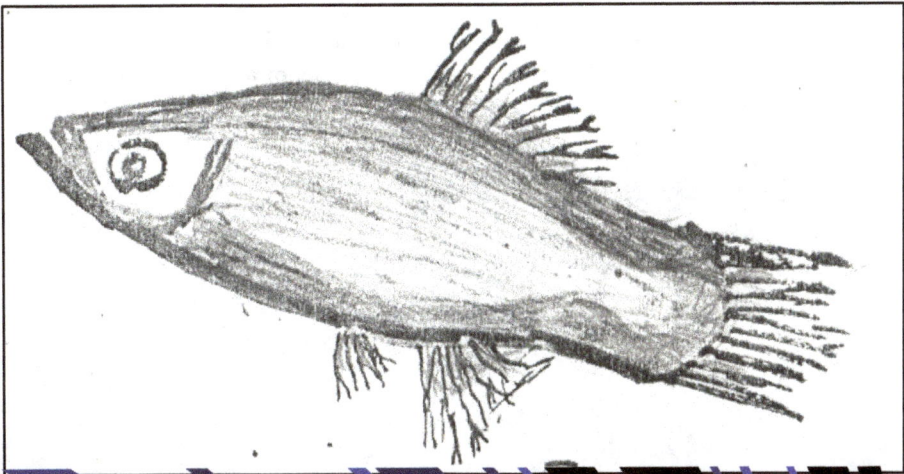

Figure 9.2: Fin Rot

Figure 9.3: Tuberculosis with Tumor Development

Figure 9.4: Black Widow Showing Typical "Ick" Sign, of a Folded Dorsal Fin, this Sign Often Appears in Early Stage

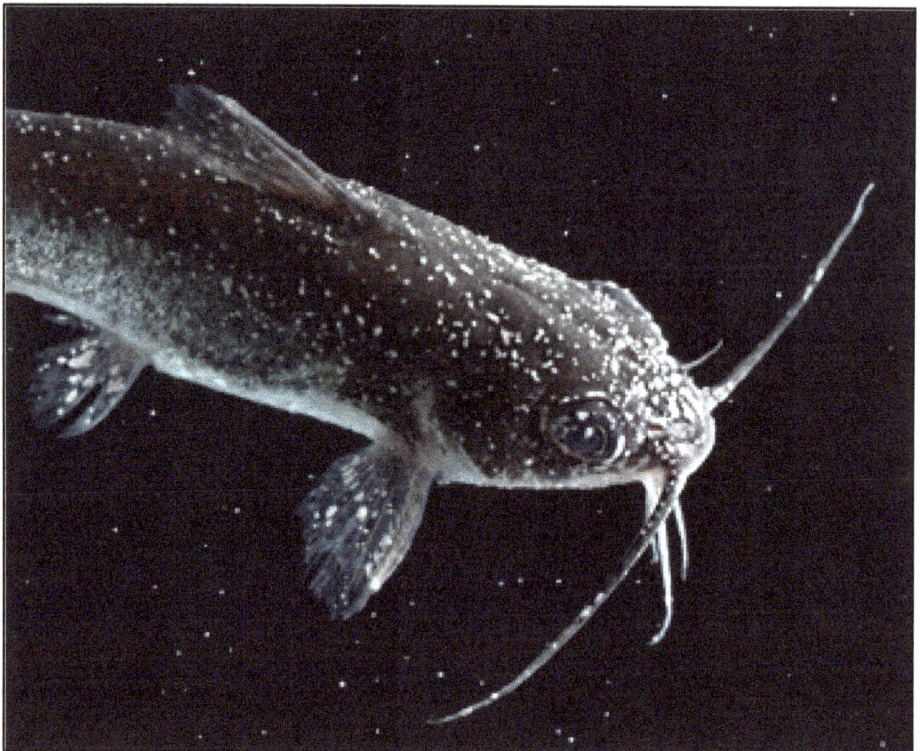

Figure 9.5: A Catfish Heavy Infested with White Spot

Figure 9.6: Discus Fish Infested by Costia

Figure 9.7: Fish Affected with *Lernaea*–Anchor Worm

Symptoms

These entire three protozoan cause the symptom for the skin disease. Dull colour, fraying of fins, weakness and damage in gills appears. A gray-white film of excess mucus develops over the body and is especially noticeable over the eyes or areas of darkened pigmentation of the skin. Badly affected fish become listless and lie on the bottom, occasionally scratching themselves against rocks. Secondary bacterial infection is common.

Treatment

Initially treat the fish with malachite green. If no improvement is observed within 5-7 days, carry a 50 per cent water change and use an antiparasitic treatment with a formalin (15-25 mg/liter, continuous bath for several days) or organophosphorus insecticides like metriphonate (0.25-0.40 mg/liter, continuous bath for 7-10 days, may need repeating). Use proprietary treatment for protozoan. *Chilodonell* is resistant so use formalin treatment. Use salt 3 per cent for salt bath with 30 gm of normal salt per liter of water.

(5) Velvet Disease (*Amyloodinium Ocellatum*) (Figure 9.15)

Very tiny parasite creates fine gold dust on skin. The roots penetrate skin and often attack gills. It also possesses pigments like plants. Disease caused by protozoan *Amyloodinium Ocellatum.*

Symptoms

Clamped fins can be seen. Respiratory distress. Fish is breathing hard. Yellow to light brown "dust" on body are appeared.

Treatment

Use proprietary treatment and remove any activated carbon from filtration during medication. It may help often to darken the tank during treatment to avoid use of light for energy production by organism. Give bath to fish in Copper sulfate at concentrations ranging from 18 mg/lt. up to 0.25 mg/lt.

(6) Finrot (Figure 9.2)

Disease usually caused by various bacteria such as Aeromonas, Pseudomonas and Myxobacteria. Fin nipping by other fish or damage by careless netting especially poor water conditions are major factors for spreading of this disease.

Symptoms

Fish is found with split, ragged or stumpy fins, often with a white edge to them.

Treatment

Carry out a general cleaning of aquarium proprietary course will help to speedy recovery. $CuSO_4$ 1:200 for 1 to 2 minutes may be given several treatments at interval of 24 hours is to be give. Chlortetracycline 10-20 mg/liter continues bath for up to 5 days. Oxytetracycline hydrochloride 20-100 mg/liter, continues bath for up to 5 days.

Caution

Tetracyclines are photo sensitive–turn lights off during treatment–better still cover the whole tank with a blanket.

(7) Columnaris: Mouth Fungus (*Flexibacter columnaris*)

Livebearers are susceptible to this particularly. Black mollies show white cotton patches around the mouth skin. Though called mouth fungus it is not fungus at all, but a slime bacteria that joins with other in colony to form threads. Mouth Fungus is so called because it looks like a fungus attack of the mouth. It is actually caused from the bacterium Chondrococcus columnaris as well as *Flexibacter columnaris*. It shows up first as a gray or white line around the lips and later as short tufts sprouting from the mouth like fungus. The toxins produced and the inability to eat will be fatal unless treated at an early stage.

Treatment

Propriety treatments usually are successful. It can be treated with range of antibiotics, but must be used in separate tank. Do not use net for transfer of fish. Use plastic bags. Penicillin at 10,000 units per liter is a very effective treatment. Treat with a second dose in two days. Or use chloromycetin, 10 to 20 mg per liter, with a second dose in two days. At early stages (only external infection) a general aquarium antibacterial or phenoxyethanol (100 mg/liter, continuous bath for at least 7 days) based remedy can be used.

(8) Septicaemia (Blood Poisoning)

Bacteria after entering the blood vessels spread through out the tissues. Damaged tissue causes leakage of fluid into the abdomen, this is seen as dropsy. Skin and base of fins becomes reddish.

Treatment

In larger fishes antibiotics may be given by injection. For smaller fishes antibiotics may be given as food in water.

(9) Dropsy

It means fluid in abdomen. It can be examined by dissection only. Mostly this happens due to severe infections of different parasites. Causal pathogen(s) not certain– possibly multiple fungal/bacterial or viral infection. It could be due to a metabolic or nutritional disorder.

Symptoms

Fish's body bloats out (as though full of roe) and, viewed from above, scales stand away from body producing a pineapple-like appearance. Ulcers on body, pale gills and a "pop eye" appearance are also common.

Treatment

As soon as abdominal swelling is noted isolate the fish and treat with a broad spectrum antibiotic. Use oxytetracycline (20-100 mg/litre; five days bath.Tetracycline hydrochloride (40-100 mg/litre; five days bath. Increase aeration during treatment. Do not use monocycline a third time in a raw. Caution: tetracyclines are photo sensitive–turn lights off during treatment. A proprietary treatment with antibacterial compound nifurpirinol is to be give. Since common cause is bacterial, treatment with suitable antibiotics on veterinary prescriptions is required.

(10) Tuberculosis (*Mycobacterium* spp.) (Figure 9.3)

Even after normal uptake of food loss of weight in fish is observed. Sometimes it developed nodules under the skin also. In some species behind the eyes it causes 'pop-eye'. Once disease is diagnosed it is essential to take great care with hygiene. Sensible precautions include washing your hands thoroughly after contact with the water.

Treatment

It requires proper specialized veterinary advice and antibiotic treatment to stand any chance of success. Fish positively identified may rarely respond to drug, but fish which come in contact with this should be treated via water. Such fish should be removed in hospital tank and disinfect the infected tank. Never allow infected fish to die in tank.

(11) Tail Rot Fungus (*Saprologina*)

It is the commonest fungus infection of aquarium fishes. Disease caused by various species of aquatic fungi, including Saprolegnia and Achlya.

It looks like tufts of dirty green cotton wool attached to the body of fish due to algal growth on fungus. Fungal infection is always secondary to another problem.

Symptoms

Gray, brown or white cotton-wool-like growths or tufts are appeared on the skin and fish of freshwater and brackish fish. Begins as a small patch but can develop and quickly kill the fish.

Treatment

Transfer the fish to clean water. Treat with Malachite green: Oxalate crystals zinc free (1: 10,000) for 2 minutes. Methylene blue can also be used; 2 mg/liter, continuous bath for several days–repeat if needed.

(12) Attach of Parasite *Trichodina* (Figure 9.10)

Trichodina sp: Circular in shape, with a band of cilia around the circumference. Size 40–60 microns in diameter, and have a rotating movement. They have a disc which has many teeth, which they use to "hook" themselves onto a host. The parasites most frequently attack the gills of the infected fish, which causes in heavy infestations, great difficulty in breathing, so that they come to the surface in a desperate attempt to get sufficient oxygen.

Typical Signs of Infection

Behaviour

Lethargy, and scratching against any suitable object. Fish is generally found breathing at the surface, or just stationary "hanging" at the surface.

Fins

Fins often become clamped or folded.

Body

The body will manifest darker colours than normal.

Different Types of Parasites

Figure 9.8: Argulus (*Gyrodactylus*)   **Figure 9.9: Skin Flukes**

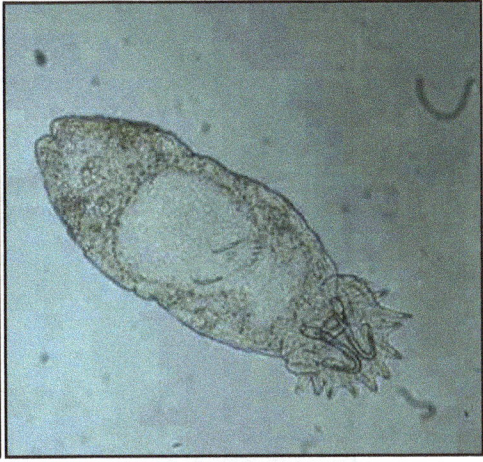

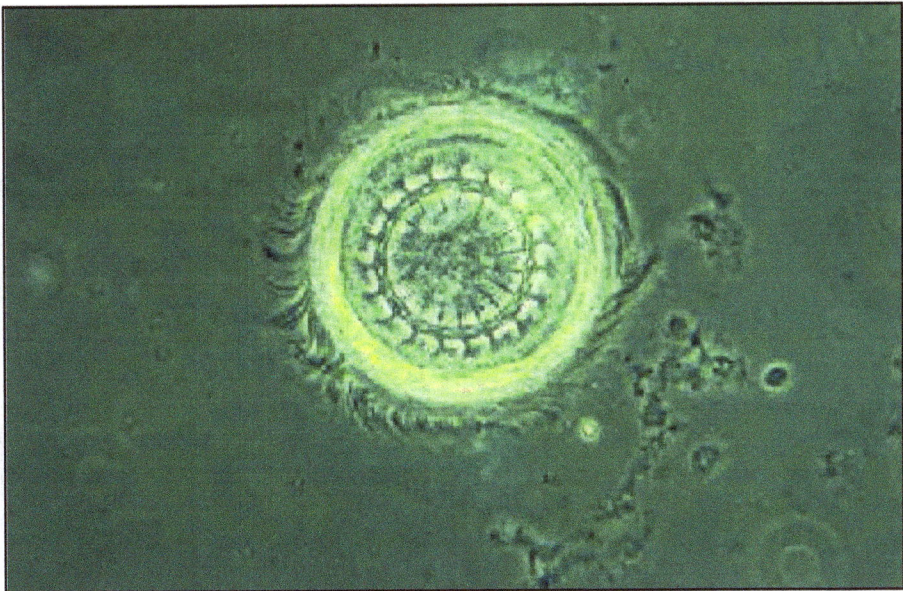

Figure 9.10: *Trichodina* (Slimy skin)

Gills

Gill examination will show large numbers of the organisms, and an excess of mucus.

Skin (Smear)

A pale bluish slime is often noticed which covers the skin; this is typically blotchy in nature.

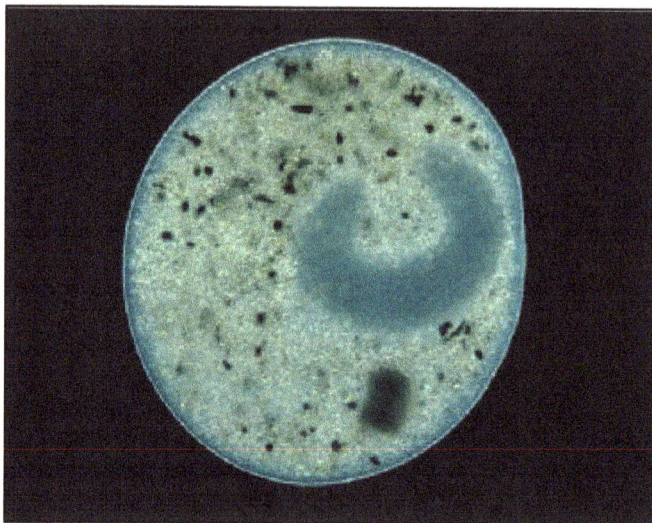

Figure 9.11: White Spot (*Ichthyophthirius multifiliis*)

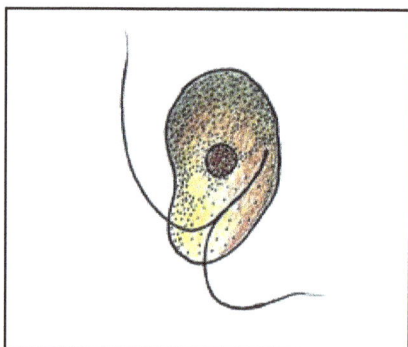

Figure 9.12: Costia

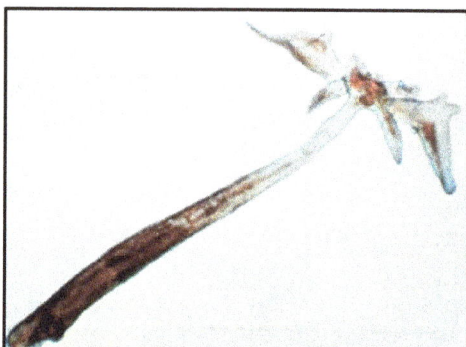

Figure 9.13: *Lernaea*–Anchor Worm

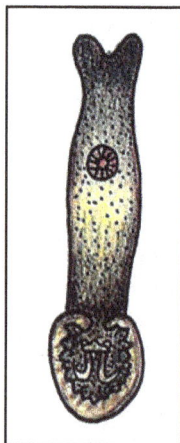

Figure 9.14:
Gyrodactylus–
The Gill Fluke

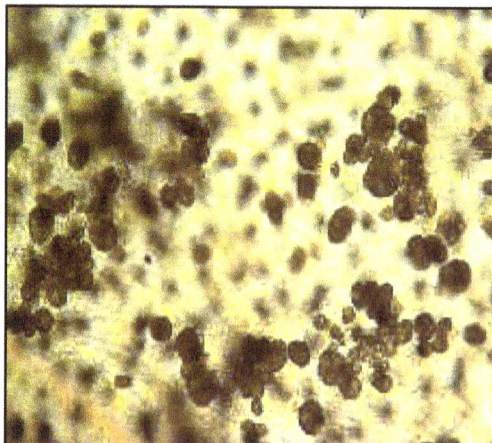

Figure 9.15: *Amyloodinium ocellatum*

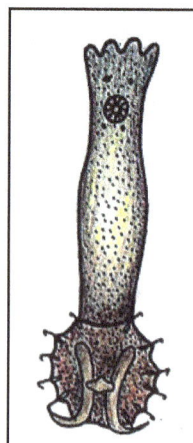

Figure 9.16:
Dactylogyrus–
The Gill Fluke

Treatment

Malachite green 0.2-0.25 ppm but it is badly tolerated by many of the Tetras, and especially so, by scale less fish such as Elephant noses and Clown loaches.

Salt baths. In a 1 per cent solution for about 30 minutes, this repeated for a couple of days.

Formalin (37 per cent–40 per cent) 250 ppm for about an hour, or 100 ppm for 3 hours. Long term baths 15-20 ppm.

Acriflavin Use at 50 ppm as a bath for about 2-4 days.

Methylene blue 100 ppm as a bath 2-4 days.

(13) Attack of Costia (A Parasite with Flagella) (Figure 9.12)

Costia is a minute flagellate with 3-4 flagella. It affects both the skin and gills of fish. Fish suffering infestations exhibit the classic symptoms of lethargy, clamped fins, rubbing and flashing and the skin can take on a grey white opaqueness.

Symptoms

Fish shows milky cloudiness on skin.

Treatment

The best treatment is with copper at 0.2 mg per liter (0.2 ppm) to be repeated once in a few days if necessary.

Acriflavine may be used instead at 0.2 per cent solution (1 ml per liter). As acriflavine can possibly sterilize fish and copper can lead to poisoning, the water should be gradually changed after a cure has been effected.

Salt bath 3 per cent solution

Raising the water temperature to 80°–83° F for a few days has also been effective.

Once the disease is identified, Table 9.1 will give you the method of treatment.

Hospital Tank

With some diseases it is better to remove the fish to a hospital tank. In addition, is isolation in a separate aquarium is always needed where the medication would disturb the function of the bacterial filter system. These substances include Methylene blue, most antibacterial treatments and antibiotics.

Suitable Heater

Suitable heaters are easily adjustable thermostat. A "COCCON" of plastic mesh around the heater will prevent fish resting too close to the heating element.

Simple Filtration

Simple filtration means of a box foam or internal power filter. Filters should not contain activated carbon as this removes many dedications form the water.

Table 9.1: Treatment of Tropical Fish Diseases

Sl.No.	Disease	Sign	Treatment
1.	Constipation	Loss of appetite, long string of face	Feed live fish food worms etc.
2.	Dropsy	Bloating of body scales protrude.	Tap carefully with a syringe aeromycin (250 mg per gallon) sometime helps.
3.	Eye Fungus	Eye covered by whitish sumps, later cotton like growth appears.	Paint 1 per cent silver nitrate on eye, and body with 1 per cent potassium dichromate.
4.	Fish louse	Not a disease but an animal parasite, sucking blood from the skin.	Touch the parasite with dry salt, disinfect would after words with mercurochrome or hydrogen peroxide.
5.	Flukes	Also caused by animal parasites. Fish rubs itself against any hard object paint skin become slimy small blood spots on body.	Treat with 5 drops of 5 per cent methylene blue per gallon of water or 1:100 formalin.
6.	Gill disease	Slime and bits of debris protrude from the gill opening, gill darkened, pale in later stages fungus may set in.	Aueromycin (50 mg per gallon) for 2–3 days.
7.	Ich	Minute white spots on body.	Heat water to 86º F + salt (two spoonful per gallon) or quinine (50 mgs per gallon)
8.	Mouth Fungus	Not a true fungus infection. Cotton growth in mouth	Terramycin or Aueromycin or Sultediagine (50 mg per gallon)
9.	Neon disease	Akin to human T.B. body wastes away. Two pale yellow dots appear on canal peduncle. The neon blue line fades in colour.	No proved cure. Try 500 mgs Terramycin + 500 gms Sueromycin for standard 15 gallon tank.
10.	Rewhirling disease	Fish loses balance and whirl about dizzily by protozoan parasite.	No known cure.
11.	Leechas	Animal parasites sucking blood	2.5 per cent salt bath for ½ hour or paint wound with mercuro chrome.
12.	Rope Eyes	Eyes bulges out	Caused by too much oxygen in water. Remove out aquarium to shady place. Stop aeration.
13.	Saprolegnia (Fungus)	Cotton growths on body usually a secondary infection on a wound or injury.	Paint with 5.5 per cent Methylene blue or Malachite green. Keep egg in gms of 5 per cent Methylene blue per gallon.
14.	Slimy skin disease (Costains)	Body get coated with bluish grey alime	3 per cent salt bath, followed by 1: 4000 formalin for 1 hour. Sterilize aquarium and plants.
15.	Spottiness skin	Affect labyrinthine fish's whitish or body patches on skin and fins.	Heat (90º F) + salt for 2 hours.
16.	Swim bladder trouble	Fish has difficulty in maintaining balance or level	May be caused by poor diet or dudden change in temperature. No cure.

Contd...

Table 9.1–Contd...

Sl.No.	Disease	Sign	Treatment
17.	Tail rot or fin rot	Tail and fins rot away leaving a white margin.	Cut off infected part of fins. Treat with suermycin (100 mgms per gallon) + 2 drops of 5 per cent methylene blue per gallon.
18.	Tuberculosis	Sluggishness, loss of appetite body waster away.	Siremtomycine (10 gm per gallon) may help. Avoid heavy crowding.
19.	Velvet	Resembles itch but has smaller white spots giving velvety appearance in reflected light.	Copper sulphate (1: 500,000) is the best remedy or 2 drops of 5 per cent methylene blue per gallon. Keep tank in dark.
20.	Dermatomy- cosis	It is caused by virous member of family *Saprolegnianacae*. Appears as gray woolly out growth of fungal thread.	Isolating fish and placing in a bath of Methylene blue, Malachite green, Acriflavin, Potassium dichromate, Ozone treatment of water for a short period each day during ten days.
21.	Pox	Appears as cloudy white spots or growths on the skin.	Acriflavin bath 2–10 ppm for 12 hours every day for one week. Or Nitrofurazone bath 15 ppm for 5–8 days.

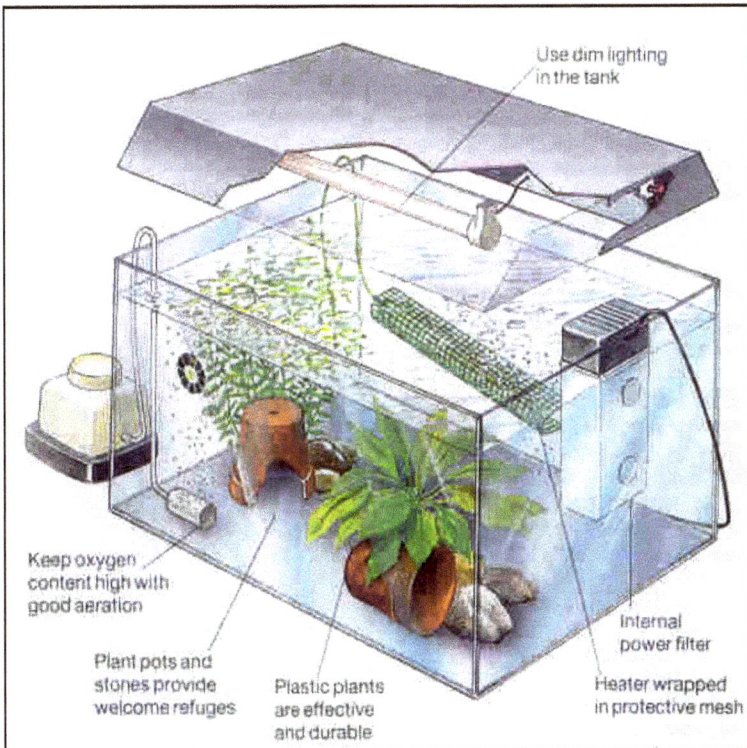

Figure 9.17: Basic Hospital Tank

Adequate Aeration

Adequate aeration since many treatments reduce the oxygen carrying capacity of the water.

Light

Dim light because some treatments are neutralized by light and others sensitize the fish to light, causing skin diseases.

Plastic Plants

Plastic plants should be provided to give sense of security. Real plants may be killed by treatments. Shelters such as rocks or pots may be provided.

Suitable Water

Water of hospital tank should be as similar as possible to water from the tank in which the fishes normally live.

Once a fish has been successfully treated, acclimatize it over a period of few days by gradually replacing water in the hospital tank with tank water. Then return the fish to the main tank by driving it into a plastic bag, but not by netting it.

Chapter 10
Aquarium Management

Introduction

Proper scientific and practical knowledge is necessary to select tank, plants and fishes for an aquarium. For successful fish keeping, different water conditions like temperature, oxygen, hardness, pH and other factors like lighting, feeding, planting etc should be taken in to account. Fish keeping would become an easy and encouraging hobby if you have a perfect understanding of the scientific management of the aquarium. The fish tank is the basic equipment for all home aquaria. Tanks are available in various shape and sizes. Different types of commonly available aquarium tanks and their capacity are given in Table 10.1.

Table 10.1: Showing Size of Panes of Glass Required, Glass Thickness, Water Capacity and Number of Fish can be Kept

Tank Dimensions (Inches)			Tank Dimensions (cms)			Glass Thickness (mm)	Surface Area	Water Capacity (Liter)	Weight of Water (kg)	Maximum Fish Capacity (Body Size) (cm)
Length	Width	Depth	Length	Width	Depth					
18	10	10	45	25	25	4	1125 cm²	27.3	27	38
18	15	12	60	38	30	6	1350 cm²	45	45	45
24	12	12	60	30	30	6	1800 cm²	54.6	54	60
24	15	12	60	38	30	8	1800 cm²	68	68	60
36	15	12	90	38	30	10	2280 cm²	104	104	90
48	15	12	120	38	30	12	1300 cm²	136	136	120

Beginners are advised to start with 60 × 30 × 30 cm size tank. It is better to keep the tank away from direct sunlight to avoid over heating and excessive algal growth.

Instead of the window, a place near the window preferably getting northern light may be selected. It is advisable to decorate the tanks after placing it to its permanent position. Care should be taken not to move the tank with water as this may lead to serve leakages.

Lights can be mounted in the aquarium cover the hood or reflector. Lamps installed at the front part of the reflector are more ideal as they produce shadows away from viewer. According to Mills (1984) a 30 cm long aquarium tank requires 40 watts (tungsten lamp) or 20 watts (Fluorescent lamp) light. Fluorescent lighting was found to be more successful and economical. The illuminating sources should not touch the water or the glass. It is suggested that an aquarium should receive 8 to 10 hours of illumination and should not receive more than two hours direct sunlight.

Most of the tropical fishes prefer temperature between 22 and 25°C. However, in some regions of tropical countries, it is not possible to get uniform temperature in all the seasons.

Good quality of water is the most important of the aquarium. Any type of the pure natural water can be used for the keeping aquarium fishes. Tap water for domestic purposes is usually disinfected by chlorine, which is harmful to fish. The effect of chlorine can be neutralized by exposure to sunlight and to strong aeration for 2 to 3 days or by adding hypo solution (Sodium thiosulphate).

Hardness of water is mainly due to dissolved salts, particularly calcium and Magnesium. Hardness, which can be removed by boiling, is usually referred as temporary hardness.

Neutral water has a pH of 7. Regarding pH requirement by each species of fish varies and it has its own preference.

The water of a living aquarium is likely to get polluted by waste products from different sources like excretory products of fish, dead plants and animals, uneaten food etc. Disintegration of nitrogenous waste material release large amount of ammonia, the concentration of which beyond 0.5 mg/liter is toxic to fish. Hence, it is necessary to oxidize the highly toxic ammonia to nitrite and then to nitrate, which is non-toxic and may be absorbed by plants as their nutrients.

Different types of filters both internal and external are available in aquarium shops to purify aquarium water. Unlike the box type filters, biological filters rely on bacteria to break down toxic ammonia to nitrite and then to less toxic nitrates.

Besides producing oxygen through the process of photosynthesis plants act as purifiers and beautifiers in an aquarium. They provide food, shelter and spawning ground for several fishes, different plants can be arranged in aquarium. The details of different aquarium plants are given in other chapters of this book.

Both natural/live and artificial/prepared food are use to feed freshwater aquarium fishes. Suitable food for different fishes are given with respective species in this book may be selected. It is advisable to feed the fishes twice a day, early in the morning and in the evening. The most important thing to remember about feeding is that over feeding is the single major cause of fish mortality and pollution in aquaria. It must be remembered that the fishes require feed for maintenance and performing metabolic activities and not for temperature regulation. *"The wisest motto in fish feeding is that hungry fish are healthy fish."* Not too much and not too little is ideal food.

Before introducing fishes into the aquarium tank, it is better to keep them in a separate tank for few days. This is usually done to observe the fish for the occurrence of diseases, if any, if symptoms of disease occur, they may be treated with suitable therapeutical agents. Details of identifying diseases and its treatment are given in this book in other chapter. Prior to releasing the fishes in to the aquarium tank, the polythene bag or container containing the fishes should be kept floating in the aquarium tank in order to acclimatize the fishes to the temperature of the tank. Before stocking one should have an idea about the number of fishes that could be stocked in an aquarium tank (Holding capacity). The holding capacity of tank depends mainly on the surface area of the tank, its oxygen content and size of the fish. Details of holding capacity of different size is given in Table 10.1. There is no definite calculation to show the exact number of fishes that could be stocked in aquarium. However, according to Mills (1989) it is practical to allow 75 cm^2 water for every 2.5 cm of fish body length (standard length). For example, an aquarium tank with a water surface area of 2700 cm^2 will support a 90 cm fish *i.e.* 10 fishes of 9 cm length or 15 fishes of 6 cm length. (Alappat and Biju Kumar,1997).

Under usual conditions, the bubbles released by aerators serve to stir the water. The usual method of aeration is to release bubbles through porous stones, kept at the bottom of the tank. Carborandum stones give the finest bubbles, but they also need powerful air pressure. All stone tend to clog, especially when not used with pressure. Therefore, they should be removed dried and reset. Fine rubber or plastic tubing should be used to lead air from pump to stones. It is suggested that small bubbles of an average diameter of about 1/12 cm, an aerating stone delivering 32 cm^3 of air/meter is adequate in a 70 liter aquarium.

The fishes and plants have symbiotic relationship and they contribute in the substance of each other. Before putting new plants in the aquarium one should ensure that they are clean and free from encrustation of germs. To do this, rinse the plant and fish with clean water and then give bath treatment of Potassium permanganate solution.

General Guideline for Fish Keeping

The following general guidelines may help the aquarists/entrepreneur in deciding suitability of fish to his purpose of aquarium fish keeping:

1. Small and social fish are ideal for a medium sized home/community aquarium.

2. Small fish in groups of 2 (male and female) or 6 are very suitable for small home aquarium, if the selected fish species dislike other fish species.

3. Small to medium size fish may be good to look at when kept as lone occupier in small but aesthetically decorated home aquarium or in separate compartments of a medium sized aquarium. If the selected fish species is very expensive to buy or rare in supply or is a fighter or a predator or just very quarrelsome by habit, or an aggressive male in breeding season.

4. Small or large fish, educative, but not necessarily ornamental, should be kept in school or public aquaria.

5. Any fish, small or large ornamental or otherwise, educative or of interest in zoological or biomedical research, may be kept for supply on order or for sale at dealer's counter for supply to laboratories and research institutes on demand.

6. Any fish which is attractive but inexpensive may be desirable, if the fish keeping is undertaken by a beginner hobbyist for the sake of learning the art of fish keeping.

(Source: Srivastava, C.B.L., 2002)

Guideline for Professionals

1. The fish species is easy to breed for developing bewildering varieties by way of inbreeding, cross-breeding (Hybridisation), selective breeding or transgenics a (a technique in genetic engineering involving introduction of new genes).

2. The fish species has the 'preferences' on account of being a rare species, threatened species, an exotic species or a 'new' species.

3. The fish species display sexual dimorphism for easy sexing of pairs–often in great demand by hobbyist, like other pets.

4. The aquarium fish must possess the following qualities,

   (*i*) It should be easy to bread and keep.

   (*ii*) It should be freely cheaply available.

   (*iii*) It should be less demanding as to its life style, habit or habitat.

   (*iv*) It should be hardy to withstand natural conditions and stresses offered in captivity and due to changes in water quality.

Some Tips for Maintaining Healthy Aquarium

Aquarium should be properly maintained in order to keep it balanced and healthy. The following tips should be kept in mine for maintaining a healthy aquarium.

1. Temperature of water should be checked and regulated daily in areas where there is drastic variation in daily temperature.

2. If you notice an oil film on the surface, remove it immediately since this prevents atmospheric oxygen from dissolving in to the water. Spreading a sheet of newspaper on the surface and towing it gently can remove the oil.

3. Growth of algae on aquarium glass could be prevented by periodically cleaning the glass panes using a razorblade scrapper or magnetic algae cleaners. Application of common salt (1 teaspoonful per 10 liter water) is supposed to be efficient in controlling the growth of this algae.

4. The working conditions of aerator, filter and heater should be checked periodically.

5. Avoid sudden switching off and on the aquarium light since this may directly interfere with the normal behaviour of fishes.

6. In order to maintain stable water conditions, partial water change. (10–25 per cent of the total water content)

7. Remove the uneaten food and waste products periodically. It can be done either by using a small hose or by using vacuum cleaner. The suction head of the hose should be at least 1 cm above the bottom in order to avoid the removal of gravel or sand.

Management Points for Aquarium Keeping

To Find Out Size and Capacity

An aquarium of 50 × 30 × 30 cm can be comfortably house six to eight 25–30 mm gold fishes or guppy. This would be hold 1 cm ft. of water with a surface area of one square foot. To find out how much water an aquarium would hold following formula suggested by Kuldip Kumar (2004) is used.

$$\frac{\text{Length in inches} \times \text{width in inches}}{1728} = \text{Capacity in Cubic–foot.}$$

Aeration

Under usual conditions, the bubbles released by aerators serve to stir the water. The usual method of aeration is to release bubbles through porous stones kept at the bottom of the tank. Carborandum stones give the finest bubbles, but they also need powerful air pressure. All stones tend to clog, especially when not used with pressure. Therefore, they should be removed dried and reset. Fine rubber or plastic tubing should be used to lead air from pump to the stones. It is suggested that small bubbles of an average diameter of about 1/12 cm, an aerating stone delivering 32 cm^3 of air/meter is adequate in a 70 lit aquarium.

Light

It is suggested that an aquarium should receive 8–10 hours of illumination and should not receive more than two hours of direct sunlight.

Artificial Light: Some Tips

1. Too much light is bad as it may lead to undesirable growth of green algae spoiling visibility.

2. Too little light is also bad as it may cause appearance of another type of algae, the brown algae.

3. Illumination when necessary must come from above rather than from any one side. The latter generally may obscure fish viewing owing to what is called, swimming–"at an angle" response to direction of light source.

4. Natural (diffused) light is essential for fish. Artificial light is desirable for giving the aquarist the advantage of freedom of placing the aquarium any where he/she likes in a closed room and to meet the exhibition requirement of fish for the onlookers.

Plantation

The fishes and plants have symbiotic relationship and they contribute in the sustenance of each other. Before putting new plants in the aquarium one should ensure that they are clean and free from encrustation of germs. To do this rinse plants with five per cent KMnO$_4$ solution.

Table 10.2: Economics of Ornamental Fish Breeding on Small Scale

Sl.No.	Details	No	Unit cost	Total Cost
I	**Capital Cist**			
1.	Glass Tanks			
	48" × 18" × 18"	6 Nos.	Rs. 600 each	Rs 3600.00
	24 " × 12" × 12"	12 Nos.	Rs. 200 each	Rs 2400.00
2.	Stands			
	48" × 18" × 18"	6 Nos.	Rs 500 each	Rs 3000.00
	24" × 12" × 12"	12 Nos.	Rs 300 each	Rs 3600.00
3.	Cement cisterns of 500 lit.	6 Nos	Rs 500 each	Rs 3000.00
4.	Air Compressor	1 No	Rs 2000 each	Rs 2000.00
5.	Aeration accessories, Plankton net, scoop net, buckets, mugs etc.			Rs 1000.00
	Total			**Rs18,600.00**
II	Variable Cost–A			
1.	Cost of brooders 200 pairs			Rs 6000.00
2.	Cost of feed, manure's, chemicals			Rs 1000.00
3.	Electricity Charges			Rs 500.00
4.	Maintenance charges			Rs 800.00
	Total			**Rs 8000.00**
III	Fixed Cost–B			
1.	Interest on capital cost	@ 12 per cent		Rs 2094.00
2.	Interest on working capital	@ 14 per cent		Rs 1120.00
3.	Depreciation cost	@ 20 per cent		Rs 3360.00
IV	**Total of A + B**			**Rs 14574.00**
V	Gross income sale of 1 lakh fish	@ 50 paisa per piece		Rs 50,000.00
VI	**Net Income**			**Rs 35,426.00**

Source: Amita Saxena, 2003.

Economics

Amita Saxena (2003) has given the economics of ornamental fish breeding on small scale. The same is given in Table 10.2. According to her, initially Rs 18,600/- is to be invested as capital cost and Rs 8000/- needs as recurring cost. Rs 14,574/- is considered as interest and depreciation. Hence the gross income on sale of 1 lakh fish is Rs 35,426/-. Hence it is an economically viable business.

Water Hardness

In many literature water hardness is given using different forms like dH°, or Mg/lit CaCo$_3$ specific gravity, salinity etc. Table 10.3 gives the co-relation between dH° and Mg/lit CaCO$_3$ as well as specific gravity and salinity for common man to understand the quality of water to be used in aquarium.

Table 10.3: Water Hardness in Comparative Terms

Based on Hardness			Specific Gravity and Salinity		
dH°	Mg/lit CaCo$_3$	Considered as	Type of Water	Specific Gravity	Salinity
6	0-50	Soft	Freshwater	1.0000	1.1
3–6	50–100	Moderately soft		1.0001	1.2
6–12	100–200	Slightly hard		1.0002	1.3
12–18	200–300	Moderately hard		1.0004	1.6
18–25	300–450	Hard		1.0006	1.9
Over 25	Over 450	Very hard		1.0008	2.1
				1.0010	2.4
				1.005	7.6
				1.010	14.1
				1.015	20.6
				1.020	27.2
			Sea water	1.025	33.7
				1.030	40.2

Fish Health and Hygiene

Fish, which are very at home in the aquarium, feel very secure and so tend to be swimming confidently around the entire tank are in a state of good health. On the contrary, if they feel insecure they tend to pass most of the time hiding in a corner or behind a cover, shying away from other occupant fish–a sure sign of fish in distress.

Colour sharpness is a sure valuable guide to the state of fish's health. If colour sharpness and intensity is maintained it is indication of good health. Any sudden change or loss of colour or fish turning black is a sign of aquarium mismanagement in terms of water quality or of illness due to bad hygiene. It needs lot of experience on the part of the aquarist to ensure that change in colour pattern is not due to any of the following exceptional situations:

(a) Fear leads to colour fading

(b) Mating season intensifies colouration

(c) Colour change is not in response to changes in light conditions or background changes (blending or camouflaging with the surrounding).

(d) Newly introduced fish show initial changes in colour which are transitory in nature until fish has settled down well.

Some of the first warning symptoms of outbreak of disease or environmental mismanagement are: loss of appetite, enhanced irregular breathing, rubbing act, act of vibrating and shaking of body, clamping of fins, loss of balances, surfacing etc.

The golden rule is to at once remove the ailing fish from the aquarium tank in order to save others from the risk of contamination. "Prevention is better than cure" is a time tested adage. Now if it is found that the sick fish is not incurable, every effort must be made to save the life.

Marketing

India's share in this global aquarium fish trade is only 0.007 per cent. In term of money, it comes to Rs. 23 million or 2 crores and thirty lakhs. However, India has a domestic market of Rs. 10 crores in Ornamental fish trade and this is fast growing. MPEDA (Marine Products Export Development Authority) has estimated that India's potential for trade in aquarium fish is 5 billion US dollars which we can earn by the export of ornamental fishes. "Nabard" has visualized in 2004 the rise in our trade up to 1 per cent global trade in five years.

Export

90 per cent of Indian Ornamental fish export takes place from Kolkotta, followed by 8 per cent from Mumbai and 2 per cent from Chennai. The survey was conducted during 1998–2001 records that there are about 120 businessmen involved in the ornamental fish trade at Haat Market. Among them, about 87 are directly involved in fish and the rest with accessories.

Self Employment and Export Orientation

To encourage self employment and export financial support is provided by MPEDA by way of reimbursing 10 per cent of FOB value of export, subject to a maximum of Rs 2.00 lakh per exporter.

About 85 per cent of ornamental fish species sold in the biggest ornamental fish market of India at Kolkatta–locally called "Hatibagan Haat" are from wild collections in the rivers, springs nallas, ponds direct waters in thee state of West Bengal, Assam, Meghalaya, Arunachal and Manipur. Only 15 per cent are the exotic fishes bred in captivity. In the eastern part of Rajasthan especially around Bharatpur emphasis is on wild fishes. The nearest markets for Rajasthan are Agra, Delhi and Jaipur.

The other two centers *viz.* Mumbai and Chennai lay emphasis on breeding of exotic varieties of ornamental fishes. This is because of the ideal climatic conditions.

Ornamental fish trade is smaller proportion also exist in Himachal Pradesh that has larger water resources in the form of rapids, ponds and tanks in the hilly region. From here wild fish varieties are collected and sold in Simla, Chgandigarh, Delhi etc. (Durve, V.S., 2005).

Lighting in Aquarium

It is important to make accurate and proper lighting in aquarium. There are many types of lighting like natural light, artificial light, tungsten lamp, fluorescent tubes, spot lamps such as metal-halide and mercury vapor lamps. Advantages and disadvantages of such light are as under:

Type	Advantages	Disadvantages
Natural light	☆ Correct spectral range for all animals and plants kept in aquarium ☆ Excellent for encouraging algal growth ☆ Available free	☆ Uncontrollable and unpredictable
Artificial light	☆ Controllable	☆ Varies according to type of instrument
Tungsten lamp	☆ Cheap to install and easy to replace	☆ Poor and spectrally unbalanced light output. ☆ Run hot and have a short life. ☆ Expensive in use ☆ Dangerous near water ☆ Not Recommended.
Fluorescent tubes	☆ Cool running ☆ Inexpensive in use and long lasting ☆ Available in a range of colours ☆ Many close to natural spectrum.	☆ Relatively easy to buy and install ☆ Heavy starting gear. ☆ Performance may decline ☆ Recommended.
Spot lamps	☆ High light output making them ideal for producing dramatic effects and for use with deep tanks	☆ Expensive ☆ Light may not be totally ideal in spectral balance.

1. Ornamental fish farming is very profitable venture which requires less initial investment compared to prawn farming.
2. As the investment is less, the risk of production is also less.

Regular Maintenance Schedule of Aquarium

(A) Daily

☆ Check temperature.
☆ Check live stock.

☆ Remove protein-skimmer waste.

☆ Check all equipment is functioning properly.

☆ Remove any uneaten food.

(B) Every Other Day

☆ Top up evaporated water.

☆ Remove algae from front glass.

(C) Weekly

☆ Check residual current breaker(RCB) operation for electrical safety in the aquarium.

☆ Clean cover glasses.

☆ Remove any 'salt creep'.

☆ Add trace elements, pH buffers and vitamin supplements.

(D) Every Two Weeks

☆ Partial water changes–20 per cent of net tank content ideally..

☆ Test for ammonia, nitrite, nitrate and specific gravity in an established aquarium (this needs doing more often in a new tank).

☆ Remove any detritus from tank base or filters.

☆ Replace filter floss in canister filters.

(E) Monthly

☆ Rake through coral sand.

(F) Bi-monthly

☆ Replace any air stone, including those in a protein skimmer.

☆ Check all electrical connections.

☆ Harvest unwanted algae (more often if necessary).

☆ Change carbon.

☆ Clean protein skimmer.

☆ Rinse biological foam filters in tank water.

(G) Quarterly

☆ Clean out all canister filter hoses with hose crushes.

☆ Clean pumps and check for wear.

☆ Change air filters.

☆ Clean out all water courses on a 'systemized' aquarium.

(H) As and When Required

☆ Replace lighting tubes and bulbs according to manufacturers' instructions.

☆ Clean lighting reflectors.

☆ Record all events, test results etc. in a diary for later reference.

Ornamental Fish Breeding and its Management

Filtration

Efficient filtration is vital in all aquariums but livebearers are particularly susceptible to bacterial skin infections caused by *Flexibacter columnaris*, unless the water is well filtered, skin problems will occur.

Sieving

In sieving a proportion of the relatively large particles in the water can be removed by trapping them with glass wool or nylon nappy, or small scale, with filter paper. Nylon is preferable to glass wool, which some times fragment so that particles of glass enter the circulating water and become lodged in the gills of fishes. Sieving is of only minor importance in aquarium work.

Biological Filtration

Biological filtration systems deal with the dissolved waste products of fish, primarily ammonia. In acidic conditions, ammonia (NH_3) is ionized to ammonium (NH_4^+), which is non-toxic. In alkaline water, however, it remains as toxic ammonia. At a low level this damages the gills and at a high level it causes brain damage and death.

In the bacterial activity common to all biological filtration systems, the ammonia is taken up by *Nitrosomonas* bacteria, which live in the water of all thanks, and is converted to nitrites (NO_2^-). Another group of bacteria, called *Nitrobacter*, which live attached to gravel and other surfaces, convert the nitrites to nitrates (NO_3^-). The dissolved nitrates are much less toxic for freshwater fish, although different types of fishes differ in their tolerance to nitrate levels. In combination with the phosphate in the faeces (waste from the vegetable components of the fishes' diet), the nitrates act as a plant fertilizers used by farmers to boost crop yields.

Because of the increased nitrate and phosphate levels, algal growth may be problem if the aquarium is not well planted with mature, actively growing plants that will use up the ready supply of plant food. A certain amount of algal growth is not necessarily bad from the fishes' point of view; since most like to eat algae, but for the aquarist it impairs visibility and makes the tank look dirty and neglected.

The most effective way of controlling nitrate and phosphate levels is to make regular partial water changes. Removing approximately 25–30 per cent of the aquarium water every fortnight is adequate. Do not wait until the algae have become a problem. Once they have become established it is too late to control their growth in this way.

Replacing old plants with new young stock is another measure that will inhibit algae, because actively growing vigorous plants remove more nitrate and phosphate than do older plants in the aquarium.

Under gravel filters are designed to support a bed of gravel, which provides a large surface area on which the bacteria can grow. Water is drawn evenly down through the gravel bed to the filter plate at the base of the tank. A minimum depth of 5 cm, using 5 cm gravel, is needed above the filter plate; the ideal depth is about 7.5 cm. For the best effect the bed should be even; aqua scraping tends to disturb the water flow and draw more water and solids into shallow gravel areas where the flow is greater.

Systems have been developed which are said to reduce the high nitrate levels, which can build up in mature aquariums. These make use of anaerobic bacterial processes, i. e. those occurring in the absence of oxygen. The bacteria us nitrate as a source of oxygen to combine with the carbon present in waste matter and ultimately produce free nitrogen gas form the nitrate. This then simply diffuses out of the aquarium.

Utilization of Nitrates using Aerobic and Anaerobic Bacteria

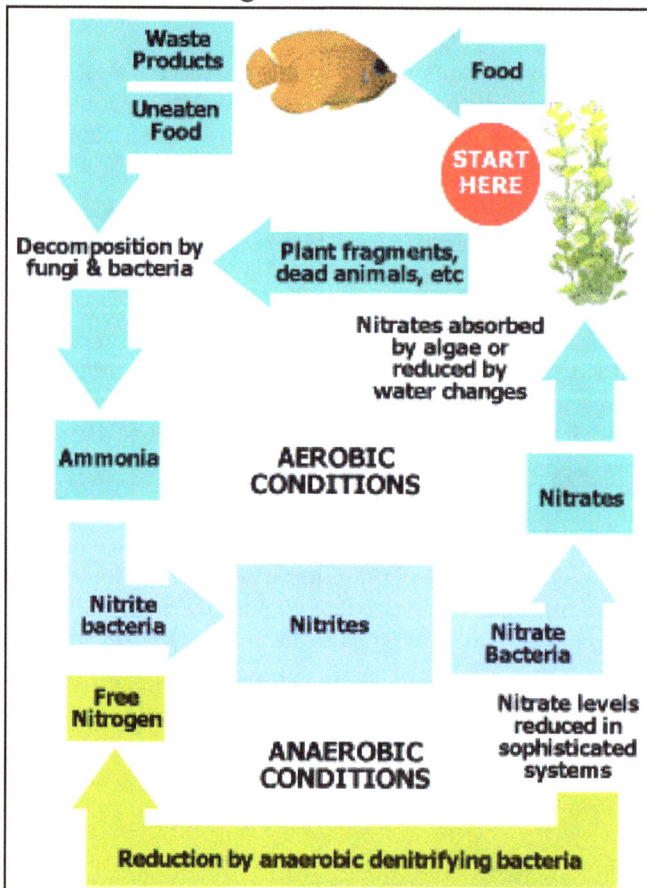

Management Against Algal Problem

During aquarium keeping some time you can face algae problem. Few following suggestions are given for curing the nuisance algae problem.

1. Feed live stock very sparingly–once a day is usually sufficient for the majority of species.
2. Do not over stock or add any new stock while slime algae persists.
3. Reduce stock if necessary.
4. Maintain good water circulation at all times, especially if reverse–flow filteration is used.
5. Do not use algal fertilizers or invertebrate food.
6. Continue to use pH buffers and trace elements.
7. Make sure all filters are efficient and working properly with the currently rated pumps.
8. Carry out proper regular maintainance and try to use high quality of water for water changes of the correct quantity every two weeks–more if necessary.
9. Do not over estimate the net capacity of the aquarium.
10. Always operate a protein skimmer and marine grade activated carbon filter.
11. Check lighting of a good quality and enough variety exists to provide a correct spectral range.
12. Do not alter the lighting period (Photoperiod) out side recommended limits.
13. Do not add higher algae until slime algae are under control.
14. Use purified water for topping up evaporated water.

Important Poins for Better Aquarium Keeping

Most important points to be kept in mind for better aquarium keeping are as under.

1. Avoid over crowding of fishes in your tank.
2. Don't keep larger fishes with smaller ones.
3. Don't overfeed your fishes, uneaten food will cause pollution.
4. Don't under-feed your fishes, they will become disease prone.
5. Provide fishes with a combination of artificial and live food.
6. Keep the filter apparatus clean and also change the aquarium water partially from time to time.
7. Don't change the water conditions of the aquarium suddenly, it becomes stressful to fishes.
8. Always buy healthy, active and brightly coloured fishes.
9. Quarantine all new additions before introducing them into your main aquarium.
10. Remove any sick fish as soon as possible.

How to Develop Ornamental Fish Farming

Although ornamental fish culture is an age old practice in India, till now it is not a popular enterprise. So, more exhibitions, training programmes, seminars and symposiams are required in order to popularize ornamental fish farming. For ornamental fish farming following suggestions may be considered.

☆ To the resource poor farmers engaged in ornamental fish culture, the concerned government should provide financial assistance in terms of long term loans with adequate subsidies.

☆ Many of the fisher folks do not know the value of ornamental fishes and they use them as food. So, a general awareness should be created among the fishing community about these much valued organisms.

☆ As ornamental fishes are of great attraction to everybody, marine/ fresh water aquaria can be set up in tourist area. They will be good source of revenue through a nominal entrance fee.

☆ Like they do in the case of prawn farming, MPEDA and other export promotion agencies should come forward to offer technical and financial assistance to the farmers who enter this virgin field.

☆ Marine Ornamental fishes fetch good prices compared to fresh water variety. But the major problem is that breeding through artificial propagation has not been standardized on a commercial scale.

☆ An ornamental fishes are to be transported live, greater importance should be given to conditioning prior to packing in order to minimize the risk of mortality due to stress.

☆ As the fisher folks are not well versed in catching marine ornamental fishes. They often unwittingly destroy coral reefs, the real habitat of these species. Therefore, suitable gear should be provided and scientific catching methods are to be developed to conserve the natural resources.

☆ In ornamental fish culture, the maximum mortality is found within 3–4 days after hatching when the yolk sac is consumed fully by the young ones and the mouth is to be formed for taking natural food. If natural food is not suitable to them, they die immediately. Therefore, live food culture is a must like in the case of prawn culture in order to avoid premature death of the fishes.

Development of Marine Aquarium Fisheries in Indian Context

A comprehensive assessment of the biodiversity of marine ornamental animal from our reef areas has not yet been made. This should be deserve primary attention. Secondly, till date no marine ornamental fisheries policy has been formulated. For this reason an organized trade of marine ornamental fish has not yet been initiated in the country. In this context, it has to be admitted that a good deal of illegal collection of marine ornamentals is in vogue in many of our reef ecosystems.

It is time to develop an organized marine aquarium fishery in India by developing proper policies and management to ensure their sustainability.

The beautiful marine coral fishes commonly kept by aquarists. The marine fish is new and exciting pass time where few really knowledgeable people exist. The distribution is restricted due to certain requirements in salinity, temperature, food etc. Marine fishes are further restricted to certain depths of oceans. | Some fishes are adapted to surface water (Anon., 1999).

Marketing of Marine Aquarium Fishes

Based on Global Marine Aquarium Database (GMAD) the annual global trade is between 20 million and 24 million number for marine ornamental fish. The annual global marine ornamental fish trade estimated at US $ 200–300 million. A total of 1471 species of fish are traded globally. Most of these species are associated with coral reefs although a relatively high number of species are associated with other habitats such as sea grass beds, mangroves and mud flats.

Among the most commonly traded families of fish, *Pomacentridae* dominates accounting 43 per cent of all marine fish trade. They are followed by species belonging to *pomacanthidae* (8 per cent), *Acanthuridae* (8 per cent), *Laridae* (6 per cent), *Gobiidae* (5 per cent), *Chaetodontidae* (4 per cent), *Callionymidae* (3 per cent), *Microdesmidae* (2 per cent), *Serrassidae* (2 per cent), and *Blennidae* (2 per cent) (G. Gopakumar, 2004).

The ornamental fish market of Kolkata (locally known as "Hatibagan Haat") is the largest wholesale market or ornamental fishes in Eastern and North Eastern Zone of India.

Control of Over Exploitation of Marine Ornamental Fish

Marine Aquarium Council (MAC) Established in 1996, has developed certification scheme that will track on animal from collector to hobbyist. This is to avoid illegal and over exploitation of marine ornamental fish from environmental area.

Chapter 11

Ornamental Fishes Available from Natural Source in Different States of India

Introduction

Different workers have given list of ornamental fishes available in different states of India. Table 11.1 gives consolidate picture of number of ornamental fish species found in respective States of India. Manoj Das and Apurba Kumar Das (2005) have given the list of 66 species of ornamental fish found in Assam (Table 11.2). Similarly Mahapatra *et al.* (2005) have given the list of 165 species available in Arunachal Pradesh, 188 species found in Assam, 137 species found in Manipur, 158 species in Meghalaya, 48 species in Mizoram, 69 species in Nagaland, 29 species in Sikkim and 121 species in Tripura (Table 11.3). Kumari and Yadav (2006) have given the list of 62 ornamental fish available in Bihar Wetland. (Table 11.4) Joe and Fofandi (2003) have given the list of 53 ornamental fish reported in Saurashtra–Gujarat. (Table 11.5) Charak and Fayaz (2005) has given the list of Potential ornamental fish culture in Jammu [Table 11.6(a) and (b)]. While Madhusoodan, Krupa B. (2003) has given the list of 105 freshwater ornamental fish as well as they have also listed 18 critically endangered fishes of Kerala [Table 11.7 (a),(b) and (c)]. Durva Vs (2005) has stated that 147 species of ornamental fishes of Rajasthan is from 26 genus (Table 11.8) Radha C. Das and Archna Sinha have given the details of 18 indigenous fishes identified as Exportable ornamental Fishes in West Bengal (Table 11.9).

Table 11.1: State Wise Wild Availability of Ornamental Fishes of India

Sl.No.	Name of State	Number of Ornamental Fish Species Reported to be Found
1.	Assam	188
2.	Arunachal Pradesh	165
3.	Bihar wet land	62
4.	Gujarat (Saurashtra)	53
5.	Jammu	51
6.	Kerala	158
7.	Manipur	137
8.	Meghalaya	156
9.	Mizoram	48
10.	Nagaland	69
11.	Sikkim	29
12.	Tripura	121
13.	Rajasthan	147
14.	West Bengal	18

Table 11.2: Ornamental Fishes of Assam

Order No.	Family No.	Species
Order No. I:	Family No. 1:	1. *Chitala chitala* (Ham-Buch)
Osteoglossiformes	Notopteridae	2. *Notopterus notopterus* (Pallas)
Order No. II:	Family No. 2:	3. *Gonialosa manmina* (Ham-Buch)
Clupeiformes	Clupeidae	4. *Gudusia chapra* (Ham-Buch)
Order No. III:	Family No. 3:	5. *Chela cachius* (Ham-Buch)
Cypriniformes	Cyprinidae	6. *C. laubuca* (Ham-Buch)
		7. *Salmostoma bacaila* (Ham-Buch)
		8. *Brachudanio rerio* (Ham-Buch)
		9. *Danio aequipinnatur* (McClelland)
		10. *D. devario* (Ham-Buch)
		11. *D. dangila* (Ham-Buch)
		12. *D. regina* (Flower)
		13. *Esomus danricus* (Ham-Buch)
		14. *Aspidoparia moror* (Ham-Buch)
		15. *Amblypharyngodon mola* (Ham-Buch)
		16. *Labeo calbasu* (Ham-Buch)
		17. *Puntius chola* (Ham-Buch)
		18. *P. conchonius* (Ham-Buch)

Contd...

Table 11.2–Contd...

Order No.	Family No.	Species
		19. *P. gelius* (Ham-Buch)
		20. *P. phutunio* (Ham-Buch)
		21. *P. sophore* (Ham-Buch)
		22. *P. terio* (Ham-Buch)
		23. *P. tictio* (Ham-Buch)
		24. *Osteobrama cotio cotio* (Ham-Buch)
		25. *Rasbora rasbora* (Ham-Buch)
		26. *Barilius barila* (Ham-Buch)
		27. *B. bendelisis* (Ham-Buch)
	Family No. 4: Balitoridae	28. *Acanthocobitis botia* (Ham-Buch)
	Family No. 5: Cobitidae	29. *Botia histrionica* (Blyth)
		30. *B. berdmorei* (Blyth)
		31. *B. dario* (Ham-Buch)
		32. *Lepidocephalus guntea* (Ham-Buch)
Order No. IV: Siluriformes	Family No. 6: Bagridae	33. *Mystus vittatus* (Bloch)
		34. *Mystus cavasius* (Ham-Buch)
		35. *Rita rita* (Ham-Buch)
	Family No. 7: Sisoridae	36. *Gagata cenla* (Ham-Buch)
		37. *Hara hara* (Ham-Buch)
		38. *Ailia coila* (Ham-Buch)
		39. *A. punctata* (Day)
		40. *Pseudotropius atherinodies* (Bloch)
	Family No. 8: Claridae	41. *Claria batrachus* (Linneaus)
	Family No. 9: Heteropneustidae	42. *Heteropneustidae fossilis* (Bloch)
	Family No. 10: Chacidae	43. *Chaca chaca* (Ham-Buch)
Order No. V: Beloniformes	Family No. 11: Belonidae	44. *Xenontodon cancila* (Ham-Buch)
Order No. VI: Cyprinodontiformes	Family No. 12: Aplocheilidae	45. *Aplocheilus panchax* (Ham-Buch)
Order No. VII: Synbranchiformes		46. *Monopterus cuchia* (Ham-Buch)
Order No. VIII: Perciformes	Family No. 13: Amassidae	47. *Chanda nama* (Ham-Buch)
		48. *Pseudambassis ranga* (Ham-Buch)
		49. *P. lala* (Ham-Buch)
		50. *P. baculis* (Ham-Buch)

Contd...

Table 11.2–Contd...

Order No.	Family No.	Species
	Family No. 14: Nandidae	51. *Badis badis* (Ham-Buch)
		52. *Nandus nandus*
	Family No 15: Gobiidae	53. *Glossogobius giuris* (Ham-Buch)
	Family No. 16: Anabantidae	54. *Anabas testudineus* (Bloch)
	Family No. 17: Belontidae	55. *Colisa fasciatus* (Schneider)
		56. *C. laila* (Ham-Buch)
		57. *C. sota* (Ham-Buch)
		58. *Ctenops nobilis* (McClelland)
	Family No. 18: Channidae	59. *Channa orientalis* (Bloch and Schneider)
		60. *C. punctatus* (Bloch)
		61. *C. barca* (Ham-Buch)
		62. *C. stewarti* (Playfair)
Order No. IX: Mastacembeliformes	Family No. 19: Mastacembelidae	63. *Macrognathus aral* (Bloch and Schneider)
		64. *M. pancalus* (Ham-Buch)
		65. *Masacembelus armatus* (Lacepede)
	Family No. 20: Tetradontiae	66. *Tetradon cutcuita* (Ham-Buch)

Source: Manoj Das and Apurba Kumar Das (2005).

Table 11.3: Ornamental Fish Species of North East and their Distribution

Sl.No.	Name of Species	Distribution							
		AR	AS	MN	ML	MZ	NL	SK	TR
1.	*Aborichthys elongatus* (Hora)	1	1	0	1	0	0	0	0
2.	*Aborichthys garoensis* (Hora)	0	0	0	1	0	0	0	0
3.	*Aborichthys kempi* (Chaudhuri)	1	0	0	1	0	1	0	0
4.	*Aborichthys tikaderi* (Barman)	1	0	0	0	0	0	0	0
5.	*Acanthocobitis botia* (Hamilton)	1	1	1	1	0	1	0	1
6.	*Acanthocobitis pavpmaceus* (McClelland)	0	1	0	0	0	0	0	0
7.	*Acanthocobitis zonaltermans* (Blyth)	1	0	1	0	0	1	0	0
8.	*Acanthopthalmus pangia* (Hamilton)	0	1	1	1	0	1	0	0
9.	*Acantopsis choirohynchos* (Blyth)	0	1	0	0	0	0	0	0
10.	*Ailia coila* (Hamilton)	1	1	0	1	0	0	0	1
11.	*Ailia punctata* (Day)	1	0	1	1	0	0	0	0
12.	*Amblyceps mangois* (Hamilton)	1	1	1	1	1	1	0	1
13.	*Amblyceps apangi* (Nath and Day)	1	1	0	1	0	0	0	1
14.	*Amblyceps arunanchalensis* (Nath and Day)	1	0	0	0	0	0	0	0

Contd...

Table 11.3–Contd...

Sl.No.	Name of Species	AR	AS	MN	ML	MZ	NL	SK	TR
					Distribution				
15.	*Amblypharyngodon mola* (Hamilton)	1	1	1	1	0	0	0	1
16.	*Anabas oliolepis* (Bleeker)	0	1	0	0	0	0	0	0
17.	*Anabas testudineus* (Bloch)	1	1	1	1	0	0	0	0
18.	*Anguilla bengalensis* (Gray and Hardwicke)	1	1	1	1	0	0	0	1
19.	*Aorichthys aor* (Hamilton)	1	1	1	0	1	0	0	1
20.	*Aorichthys seenghala* (Sykes)	1	1	1	1	0	0	0	1
21.	*Aplocheilus panchax* (Hamilton)	0	1	0	1	0	0	0	1
22.	*Apocryptes bato* (Hamilton)	0	1	0	0	0	0	0	0
23.	*Aspidoperia morar* (Hamilton)	1	1	1	0	1	0	0	1
24.	*Badis badis badis* (Hamilton)	1	1	1	1	1	1	0	0
25.	*Badis badis burmanicus* (Ahl)	0	0	0	1	0	0	0	0
26.	*Bagarius bangarius* (Hamilton)	1	1	1	1	0	0	1	1
27.	*Balitora brucei* (Gray)	0	1	1	1	0	0	0	0
28.	*Barilius barila* (Hamilton)	0	1	1	1	0	1	0	1
29.	*Barilius barna* (Hamilton)	1	1	1	1	1	0	1	1
30.	*Barilius bendelisis* (Hamilton)	1	1	1	1	1	1	1	1
31.	*Barilius bola* (Hamilton)	1	1	1	1	0	0	0	1
32.	*Barilius dogarsinghi* (Hora)	0	1	1	0	0	1	0	0
33.	*Barilius gatensis* (Valenciennes)	0	0	1	0	0	0	0	1
34.	*Barilius guttatus* (Day)	0	0	1	0	0	0	0	0
35.	*Barilius radiolatus* (Gunther)	1	1	1	0	0	0	1	1
36.	*Barilius shacra* (Hamilton)	1	1	1	1	0	0	0	1
37.	*Barilius tileo* (Hamilton)	1	1	1	1	0	0	0	1
38.	*Barilius vagra* (Hamilton)	1	1	1	0	1	0	1	1
39.	*Batasio batasio* (Hamilton)	1	1	1	1	1	0	1	1
40.	*Batasio tengana* (Hamilton)	1	1	1	1	0	0	0	0
41.	*Bengala elanga* (Hamilton)	1	1	0	0	0	0	0	1
42.	*Botia berdmorei* (Blyth)	0	1	1	0	0	0	0	0
43.	*Botia dario* (Hamilton)	1	1	1	1	1	1	0	1
44.	*Botia histrionica* (Blyth)	0	0	1	1	0	0	0	0
45.	*Botia rostrata* (Gunther)	1	1	0	1	0	0	0	1
46.	*Botia lohachata* (Chaudhari)	1	1	0	1	0	0	0	1
47.	*Brachydanio acuticephala* (Hora)	1	0	1	0	0	1	0	0
48.	*Brachydanio rerio* (Hamilton)	1	1	1	1	1	1	1	1
49.	*Catla catla* (Hamilton)	0	1	1	1	0	0	0	1

Contd...

Table 11.3–Contd...

Sl.No.	Name of Species	AR	AS	MN	ML	MZ	NL	SK	TR
50.	*Chaca chaca* (Hamilton)	0	1	0	1	0	0	0	1
51.	*Chagunius chaguino* (Hamilton)	1	1	1	1	1	0	1	1
52.	*Chagunius nicholsi* (Myers)	0	0	1	0	0	0	0	0
53.	*Chanda nama* (Hamilton)	1	1	1	1	0	0	0	1
54.	*Chandramara chandramara* (Hamilton)	1	1	0	1	0	0	0	0
55.	*Channa barca* (Hamilton)	0	1	0	1	0	0	0	1
56.	*Channa marulius* (Hamilton)	1	1	1	1	0	0	0	1
57.	*Channa orientalis* (Bloch and Schneider)	1	1	1	1	1	1	0	1
58.	*Channa punctatus* (Bloch)	1	1	1	1	0	1	0	1
59.	*Channa strewartii* (Playfair)	1	1	0	1	0	0	0	1
60.	*Channa striatus* (Bloch)	1	1	1	1	0	1	0	1
61.	*Chaudhuria indica* (Yazdani)	0	0	0	1	0	0	0	0
62.	*Chaudhuria khajuriari* (Talwar and Kundu)	0	1	0	1	0	0	0	0
63.	*Chela laubuca* (Hamilton)	1	1	1	1	0	0	0	1
64.	*Chitala chitala* (Hamilton)	0	1	0	1	0	0	0	1
65.	*Cirrhinus mrigala* (Hamilton)	1	1	1	1	0	0	0	1
66.	*Cirrhinus reba* (Hamilton)	1	1	1	1	1	0	0	1
67.	*Cloria batrachus* (Linnaeus)	1	1	1	1	0	0	1	1
68.	*Clupisoma montana* (Hora)	0	0	0	0	0	0	0	1
69.	*Colisa fasciatus* (Schneider)	1	1	1	1	0	0	0	1
70.	*Colisa labiosus* (Day)	1	1	0	0	0	0	0	0
71.	*Colisa lalia* (Hamilton)	0	1	0	0	1	1	0	0
72.	*Conta sola* (Hamilton)	0	1	1	1	0	1	0	1
73.	*Colisa conta* (Hamilton)	1	1	1	1	0	1	0	0
74.	*Crossocheilus burmanicus* (Hora)	1	1	0	0	0	0	0	0
75.	*Ctenops nobilis* (McCelland)	0	1	0	0	0	0	1	0
76.	*Danio dangila* (Hamilton)	1	0	1	1	0	1	0	0
77.	*Danio devario* (Hamilton)	1	1	1	1	1	1	0	1
78.	*Danio naganensis* (Choudhuri)	0	0	1	0	0	1	0	0
79.	*Danio aequipinnatus* (McClelland)	1	0	1	1	0	1	0	1
80.	*Danio regina* (Fower)	0	1	0	0	0	0	0	0
81.	*Erethestis montana montana* (Hora)	0	1	0	0	0	0	0	0
82.	*Erethestis pussiles* (Mull. and Tros.)	1	1	1	0	0	1	0	1
83.	*Esomus danricus* (Hamilton)	1	1	1	1	0	1	0	1
84.	*Euchiloglanis kamengensis* (Jayaram)	1	0	0	0	0	0	0	0

Contd...

Table 11.3–Contd...

Sl.No.	Name of Species	AR	AS	MN	ML	MZ	NL	SK	TR
85.	*Eutropiichthys vacha* (Hamilton)	1	1	1	1	0	0	0	1
86.	*Exostoma berdmorei* (Blyth)	1	0	0	0	0	0	0	0
87.	*Exostoma stuartii* (Hora)	1	0	1	1	0	0	0	0
88.	*Ganata cenia* (Hamilton)	1	1	1	1	1	1	0	1
89.	*Ganata gangata* (Hamilton)	0	1	0	0	0	0	0	1
90.	*Gangata sexualis* (Tilak)	0	1	0	1	0	0	0	1
91.	*Ganiolosa manmina* (Hamilton)	0	1	0	0	0	0	0	1
92.	*Garra annandalei* (Hora)	1	1	1	1	1	0	1	0
93.	*Garra gotyla gotyla* (Gray)	1	1	1	1	1	1	1	1
94.	*Garra gravelyi* (Annandalei)	0	0	1	0	0	0	0	0
95.	*Garra kempi* (Hora)	1	1	1	1	0	1	0	0
96.	*Garra lamina* (Hamilton)	1	1	1	1	1	1	1	1
97.	*Garra listanensis* (Viswanath)	1	1	1	1	0	1	0	0
98.	*Garra lissorhynchus* (Viswanath)	0	0	1	0	0	0	0	0
99.	*Garra manipurensis* (Viswanath and Sarojalini)	0	0	1	0	0	0	0	0
100.	*Gara maClellandi* (Jerdom)	1	0	1	1	0	0	0	0
101.	*Gara naganensis* (Hora)	1	1	1	1	1	1	0	0
102.	*Gara nasuta* (McCelland)	1	1	1	1	1	0	0	0
103.	*Gara rupecula* (McCelland)	1	1	1	1	1	0	0	0
104.	*Glossogobius giuris* (Hamilton)	1	1	1	1	1	0	0	1
105.	*Glyptothorax brevipinnis* (Hora)	1	0	0	0	0	0	0	0
106.	*Glyptothorax manipurensis* (Menon)	0	0	1	0	0	0	0	0
107.	*Glyptothorax pectinopterus* (McClelland)	1	0	1	0	0	0	0	0
108.	*Glyptothorax platypogonoides* (Bleeker)	1	1	1	0	1	1	0	0
109.	*Glyptothorax telchitta* (Hamilton)	1	0	0	1	1	0	0	1
110.	*Glyptothorax annandeleri* (Hora)	1	0	0	0	0	0	0	0
111.	*Glyptothorax cavia* (Hamilton)	1	1	0	1	0	0	0	1
112.	*Glyptothorax coheni* (Ganguly, Dutta and Sen)	1	0	0	1	1	0	0	1
113.	*Glyptothorax conirostre conirostre* (Steindachner)	1	0	0	0	0	0	0	0
114.	*Glyptothorax indicus* (Talwar)	1	1	0	1	0	0	1	0
115.	*Glyptothoraxstriatus* (Blyth)	1	0	1	0	0	0	0	0
116.	*Glyptothorax trilineatus* (Blyth)	1	1	1	1	0	0	0	1
117.	*Gudusia chapra* (Hamilton)	1	1	1	1	0	0	0	1

Contd...

Table 11.3–Contd...

Sl.No.	Name of Species	AR	AS	MN	ML	MZ	NL	SK	TR
					Distribution				
118.	*Hara hara* (Hamilton)	1	1	1	1	0	1	0	1
119.	*Hara harai* (Misra)	0	1	0	0	0	0	0	0
120.	*Hara jerdeni* (Bay)	1	1	0	0	0	0	0	0
121.	*Heteroneustes fossislis* (Bloch)	1	1	0	1	0	1	0	1
122.	*Homaloptera modesta* (Vinciguera)	0	0	1	0	0	0	0	0
123.	*Johnius coitor* (Hamilton)	0	0	0	0	0	0	0	1
124.	*Kryptoterus indicus* (Datta, Burman and Jayaram)	1	0	0	0	0	0	0	0
125.	*Labeo angra* (Hamilton)	0	1	1	0	0	0	0	0
126.	*Labeo bata* (Hamilton)	1	1	1	1	1	0	0	1
127.	*Labeo boga* (Hamilton)	1	1	1	1	1	0	0	1
128.	*Labeo calbasu* (Hamilton)	1	1	1	1	1	0	0	1
129.	*Labeo devdvi* (Hora)	1	1	0	0	0	0	0	0
130.	*Labeo dyocheilus* (Hamilton)	1	0	0	0	0	1	0	1
131.	*Labeo gonius* (Hamilton)	1	1	1	1	0	0	0	1
132.	*Labeo nandina* (Hamilton)	0	1	0	1	0	0	0	1
133.	*Labeo pangusia* (Hamilton)	1	1	1	1	1	0	1	1
134.	*Labeo rohita* (Hamilton)	1	1	1	1	0	0	0	1
135.	*Languvia ribeirai* (Hora)	1	1	0	1	0	0	0	1
136.	*Languvia shawi* (Hora)	0	1	0	1	0	0	0	1
137.	*Lepidocephalus annadalei* (Chaudhari)	1	1	1	1	0	0	0	1
138.	*Lepidocephalus berdmorei* (Blyth)	1	1	1	1	0	1	0	0
139.	*Lepidocephalus guntea* (Hamilton)	1	1	1	1	1	1	1	1
140.	*Lepidocephalus irrorata* (Hora)	0	1	1	1	0	1	0	0
141.	*Lepidocephathus menoni* (Pillai and Yazdani)	1	1	0	1	0	0	0	0
142.	*Macrognathus oral (bloch and Schneider)*	1	1	0	1	0	1	0	1
143.	*Macrognathus pancalus* (Hamilton)	1	0	1	1	0	1	0	0
144.	*Maringua hodgati* (Hamilton)	0	1	0	0	0	0	0	0
145.	*Mastocembelus armatus* (Lacepede)	1	1	1	1	1	1	0	1
146.	*Mesonoemacheilus sijuencis* (Menon)	0	1	0	1	0	0	0	0
147.	*Microphis deocata* (Hamilton)	1	1	0	0	0	0	0	0
148.	*Monopterus cuchia* (Hamilton)	1	1	1	1	0	1	0	1
149.	*Mystus bleekeri* (Day)	1	1	1	1	0	1	1	1
150.	*Mystus cavasius* (Hamilton)	1	1	1	1	0	0	0	1
151.	*Mysus menoda* (Hamilton)	0	1	0	0	0	0	0	0

Contd...

Table 11.3–Contd...

Sl.No.	Name of Species	AR	AS	MN	ML	MZ	NL	SK	TR
		Distribution							
152.	*Mysus montanus* (Jerdon)	1	1	0	1	0	0	0	0
153.	*Mysus tengara* (Hamilton)	0	1	0	0	0	0	0	0
154.	*Mysus vittatus* (Bloch)	1	1	0	1	0	0	0	1
155.	*Nandus nandus* (Hamilton)	1	1	1	1	0	0	0	1
156.	*Nangra assamensis* (Sen and Biswas)	0	1	0	0	0	0	0	0
157.	*Nangra nangra* (Hamilton)	0	1	0	0	0	0	0	1
158.	*Nemacheilus corica* (Hamilton)	1	1	0	1	0	0	0	0
159.	*Nemacheilus peguensis* (Hora)	0	0	1	0	0	0	0	0
160.	*Neoeucirrhichthys maydelli* (Banerescy)	0	1	0	0	0	0	0	0
161.	*Neolissocheilus hexastichus* (McCelland)	1	1	1	1	1	1	0	0
162.	*Neolissochelus hexagonolepis* (McCelland)	1	1	1	1	1	1	1	1
163.	*Noenemacheilus labeosus* (Kottelat)	0	1	0	0	0	0	0	1
164.	*Notopterus notopterus* (Pallas)	1	1	1	1	0	0	0	1
165.	*Olyra burmanica* (Day)	0	0	0	0	0	1	0	0
166.	*Olyra horai* (Prasad and Mukerji)	0	1	0	1	0	1	0	0
167.	*Plyra longicaudata* (Bloch)	1	1	0	1	0	1	0	0
168.	*Ompok bimacultus* (Bloch)	1	1	1	1	0	0	0	1
169.	*Ompok pabda* (Hamilton)	1	1	0	1	0	0	0	0
170.	*Ompok pabo* (Hamilton)	1	1	0	1	0	0	0	0
171.	*Ophisternon bengalensi* (McCelland)	0	0	1	0	0	0	0	0
172.	*Orwixhthya casuatis* (Hamilton)	1	1	0	1	0	0	0	0
173.	*Osphronemus goramy* (Lacepede)	0	0	0	0	0	0	0	1
174.	*Oreichthys belangeri* (Valenciennes)	0	0	1	0	0	0	0	0
175.	*Oreichthys cotico cunma* (Day)	1	1	1	1	0	0	0	1
176.	*Osteobrama cotio catio* (Hamilton)	1	0	0	0	0	0	0	0
177.	*Osteochilus neilli* (Day)	0	1	0	0	0	1	0	1
178.	*Pangasius pangasius* (Hamilton)	0	1	0	0	0	0	0	0
179.	*Pisodonophis boro* (Hamilton)	1	1	1	1	1	0	0	1
180.	*Pseudambasis baculis* (Hamilton)	0	1	0	0	0	0	0	0
181.	*Pseudambasis lala* (Hamilton)	0	1	0	0	0	0	0	0
182.	*Pseudambasis ranga* (Hamilton)	1	1	1	1	1	0	0	1
183.	*Pseudeutropius sulcutus* (Prasad and Mukhergee)	1	1	1	1	0	1	1	0
184.	*Pseudeutrous atherinoides* (Bloch)	1	1	0	1	0	0	0	1
185.	*Psilorhynchus balitora* (Hamilton)	1	1	1	1	1	1	0	1

Contd...

Table 11.3–Contd...

Sl.No.	Name of Species	Distribution							
		AR	AS	MN	ML	MZ	NL	SK	TR
186.	*Psilorhynchus gracilis* (Rainboth)	0	1	0	0	1	0	0	0
187.	*Psilorhynchus homaloptera* (Hora and Mukhargee)	1	0	0	1	0	1	0	0
188.	*Psilorhynchus sucatio* (Hamilton)	1	1	0	1	0	0	0	1
189.	*Puntius chola* (Hamilton)	1	1	1	1	0	0	0	1
190.	*Puntius clavtus* (McCelland)	1	1	1	1	0	1	1	1
191.	*Puntius conchonius* (Hamilton)	1	1	1	1	1	1	0	1
192.	*Puntius filamentosus* (Valenciennes)	0	1	0	1	0	1	0	1
193.	*Puntius fraseri* (Hora and Misra)	0	1	0	0	0	0	0	0
194.	*Puntius gelius* (Hamilton)	0	1	0	1	0	0	0	1
195.	*Puntius guganio* (Hamilton)	1	1	0	1	0	0	0	0
196.	*Puntius phutunio* (Hamilton)	0	1	1	0	0	0	0	0
197.	*Puntius sarana orphoides* (Velenciennes)	0	0	1	0	0	1	0	0
198.	*Puntius sarana sarana* (Hamilton)	1	1	1	1	0	0	1	1
199.	*Puntius shalynius* (Yazdi and Talukdar)	1	1	1	1	0	0	0	0
200.	*Puntius sophore* (Hamilton)	1	1	1	1	1	0	0	1
201.	*Puntius terio* (Hamilton)	0	1	1	1	1	0	0	1
202.	*Puntius ticto* (Hamilton)	1	1	1	1	1	1	0	1
203.	*Rasbora daniconius* (Hamilton)	1	1	1	1	0	0	0	1
204.	*Rasbora rasbora* (Hamilton)	1	1	1	1	0	1	0	1
205.	*Rhinomugil corsula* (Hamilton)	0	1	0	1	0	0	0	1
206.	*Rita rita* (Hamilton)	1	1	1	0	0	0	0	1
207.	*Salmosroma bacaila* (Hamilton)	1	1	1	1	0	0	0	1
208.	*Salmostoma clupeoides* (Bloch) *Salmostoma*	0	1	0	0	0	0	0	1
209.	*Salmostoma phulo* (Hamilton)	1	1	0	1	0	0	0	0
210.	*Schismatorhynchos nukta* (Sykes)	0	0	0	0	0	0	0	1
211.	*Schistura arunachalensis* (Menon)	1	1	0	1	0	0	0	0
212.	*Schistura beavanj* (Gunther)	1	1	0	1	0	0	0	0
213.	*Schistura vinticauda* (Blyth)	1	0	0	1	0	0	0	0
214.	*Schistura densoni dayi* (Hora)	0	0	0	1	0	0	0	0
215.	*Schistura devdevi* (Hora)	1	0	0	1	0	0	0	0
216.	*Schistura elongatus* (Sen and Nalbant)	0	0	0	1	0	0	0	0
217.	*Schistura Kangjupljilensis* (Hora)	1	1	1	0	0	1	0	0
218.	*Schistura manipurensis (Chauhan)*	1	1	1	0	1	1	0	0
219.	*Schistura multifaciatus* (Day)	1	1	1	1	1	1	1	0

Contd...

Table 11.3–Contd...

Sl.No.	Name of Species	Distribution							
		AR	AS	MN	ML	MZ	NL	SK	TR
220.	*Schistura nagaensis* (Menon)	1	0	0	0	0	1	0	0
221.	*Schistura prashadi* (Hora)	0	0	1	1	0	1	1	0
222.	*Schistura reticulofasciatus* (Singh and Banareascu)	0	1	0	1	0	0	0	0
223.	*Schisturasavona* (Hamilton)	1	1	1	1	0	0	0	0
224.	*Schistura scaturigina* (McCelland)	1	1	0	1	1	1	0	1
225.	*Schisturasikmainensis* (Hora)	1	1	1	1	0	1	0	0
226.	*Schistura singhi* (Menon)	0	0	0	0	0	1	0	0
227.	*Schistura vincigeurra* (Hora)	0	0	1	0	0	0	0	0
228.	*Schizopygopsis esocinus* (Heckle)	1	0	0	0	0	0	0	0
229.	*Schizopygopsis progastus* (McCelland)	1	1	0	0	0	0	1	0
230.	*Schizopygopsis stoliczkae* (Steindachner)	1	0	0	0	0	0	0	0
231.	*Schizothorax richardsonii* (Gray)	1	1	1	0	0	1	1	0
232.	*Securicula gora* (Hamilton)	1	1	0	1	0	0	0	0
233.	*Setipina phasa* (Hamilton)	1	1	1	1	0	0	0	0
234.	*Sicamugil cascasia* (Hamilton)	0	1	1	1	0	0	0	1
235.	*Silurus afghana* (Gunther)	1	1	0	0	0	0	0	0
236.	*Silurus berdmorei* (Blyth)	0	1	0	0	0	1	0	0
237.	*Silurus torrentis* (Kobayakawa)	1	0	0	0	0	0	0	0
238.	*Sisor rhabdophis* (Hamilton)	1	0	1	0	0	0	0	0
239.	*Somileptes gongota* (Hamilton)	1	1	0	1	0	0	0	1
240.	*Strongylura strongylura* (Van Hasselt)	0	0	0	0	1	0	1	1
241.	*Tetradon cutcutia* (Hamilton)	1	1	1	1	0	0	0	1
242.	*Tor chelynoides* (McCalland)	0	0	0	1	0	0	0	0
243.	*Tor mosal* (Hamilton)	0	1	0	0	0	0	0	0
244.	*Tor progeneius* (McCelland)	0	1	1	0	0	1	0	0
245.	*Tor putitora* (Hamilton)	1	1	1	1	0	0	1	1
246.	*Tot tor* (Hamilton)	1	1	1	1	1	1	1	1
247.	*Triplophysa grascilis* (Day)	0	0	0	0	0	0	1	0
248.	*Wallago attu* (Schneider)	1	1	1	1	0	0	0	1
249.	*Xenentodon cancila* (Hamilton)	1	1	1	1	1	0	0	1
	Total fish Species	**165**	**188**	**137**	**158**	**48**	**69**	**29**	**121**

AR: Arunachal Pradesh; AS: Assam; MN: Manipur; ML: Meghalaya; MZ: Mizoram; NL: Nagaland; SK: Sikkim; TR: Tripura; 1: Present; 0: Absent.

Source: Mahapatra, B.K., K. Vinod and B.K. Mandal, 2005.

Ornamental Fishes of Assam

Figure 11.1: *Tetradon cutcutta*

Figure 11.2: *Macrognthus aral & M. Puncalus*

Figure 11.3: *Lapidocephalus guntea*

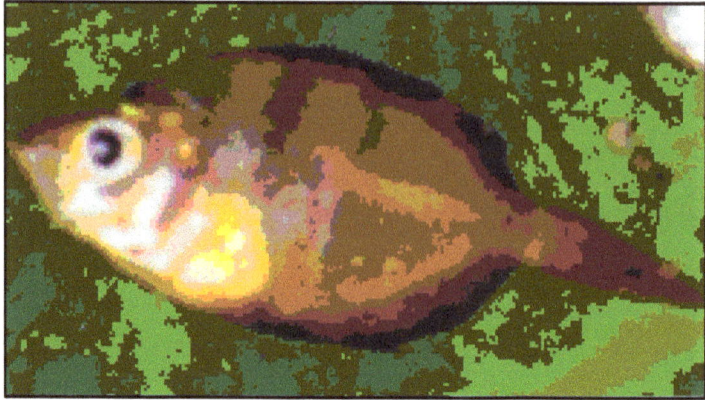

Figure 11.4: *Pseudoambassis ranga*

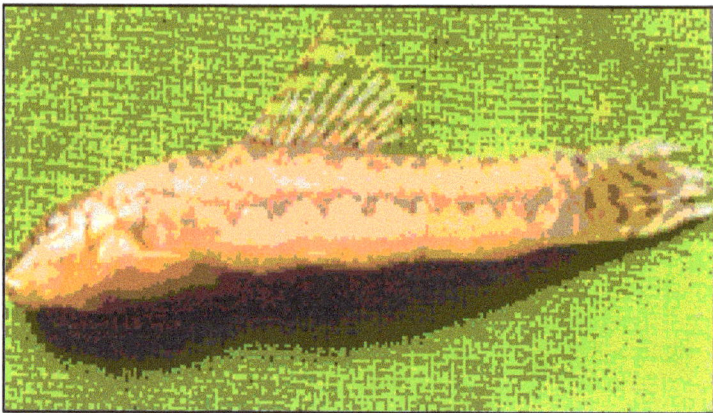

Figure 11.5: *Lepidocephalus guntea*

Freshwater Ornamental Fishes of Jammu and Kashmir

Figure 11.6: *Puntius conchonius*

Figure 11.7: *Puntius ticto*

Figure 11.8: *Nandus nandus*

Figure 11.9: *Xenetodon concila*

Figure 11.10: *Mystus vittatus*

Some Ornamental Fishes of Bihar Wetland

Figure 11.11: *Wallago attu*

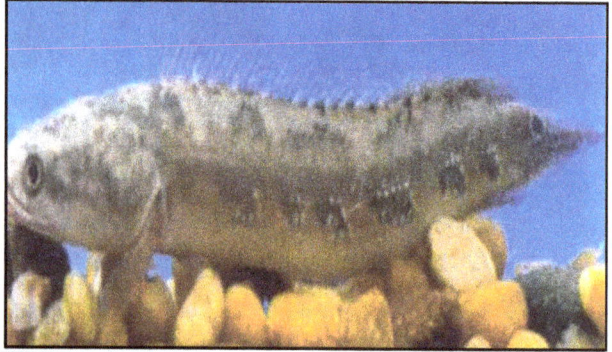

Figure 11.12: *Channa marulius*

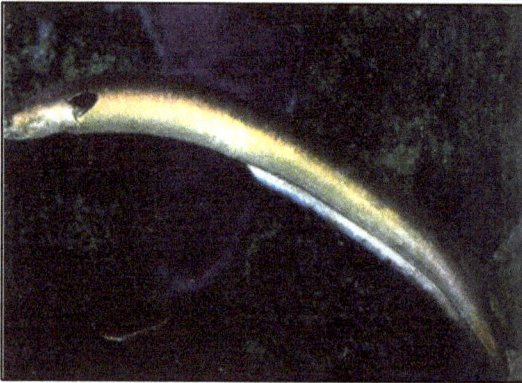

Figure 11.13: *Anguilla bengalensis*

Figure 11.14: *Ganoproktopterus thamassi*

Figure 11.15:
Ganoproktopterus curmuca

Figure 11.16: *Claria dussumeiri*

Figure 11.17: *Horiabagrus branchysoma*

Figure 11.18: *Eleotis fusca*

Figure 11.19: *Etroplus suratensis*

Figure 11.20: *Puntius sarana*

Table 11.4: Ornamental Fish of Bihar Wetland

Sl.No.	Family	Scientific Name	Local/Regional name
1.	Notopteridae	Notopterus notopterus	
2.	(Featherbacks)	Notopterus chitala	
3.	Cypernidae	Labio calbasu	
4.	(Carps and Minnows)	Puntius conchonius	Rosy barb or red barb
5.		P.gelius	Golden barb
6.		P.phutunio	Dwarf barb
7.		P. sophore	
8.		P. terio	
9.		P. ticto	Pathia
10.		Chelachela	Chela
11.		Amblypharyngodon mola	Mola
12.		Barilius barila	
13.		B. bendelisis	Zebra fish
14.		Brachydanio rcrio	Dangila danio
15.		Danio dangila	Bashpata
16.		Devario danio	
17.		Esomus danricus	
18.		Parluciosoma daniconius	Rasbora
19.		Crossocheilus latius latius	
20.		Garra gotyla gotyla	Gotyla
21.	Balitoridae	Nemachelius botia	
22.	(Loaches)	N. corica	Lotani
23.		N. scaturigina	
24.		N. beuvani	
25.		Lepidocephalus	Gunte
26.		Pagio oangia	
27.		Boita Dario	Necktie
28.		Botia lohachata	Lohachata
29.	Schilbeidae	Botiya dayi	Botye
30.	Bagridae	Pseudeutropius atherinoides	Batasi
32.		Batasio batasio	Tengara
33.		Mystus tengara	Tengara
34.		M. vittatus	Tengara
35.		Aorichthyes aor	Rita
36.		Rita rita	
37.	Siluridae	Ompak pabda	Pava

Contd...

Table 11.4–Contd...

Sl.No.	Family	Scientific Name	Local/Regional name
38.	Pangasiidae	*Pangasius pangasius*	Pungus
39.	Sisoridae	*Bagarius bagariu*	
40.		*Hara hara*	
41.		*Nangra viridescens*	
42.	Chacidae	*Chacha chacha*	
43.	Belonidae	*Xenentodon cancila*	
44.	Aplocheilida	*Aplocheilus panchax*	
45.	Ambassidae	*Chanda nama*	
46.		*Chanda ranga*	
47.	Nandidae	*Nandus nandus*	Nandus
48.	Anabantidae	*Stigmatogobius sandanundio*	Goby fish
49.	Belontiidae	*Anabus testudineus*	Koi fish
50.	(Gouramies)	*Ctenops nobilis*	Indian paradise fish
51.		*Colisa fasciata*	Stripped giant gourami
52.		*Colisa lalia*	Dwarf gourami
53.	Channidae	*Channa barca*	
54.	(Snakehead murrels)	*C. orientalis*	
55.		*C.marulius*	
56.		*C. punctatus*	
57.	Mastocembelidae	*Macrognathus aral*	Gaichi
58.		*M. pancalus*	
59.		*Mastocembelus armatus*	
60.	Soleidae	*Euryglossa pan*	
61.	Tetraodontidae	*Tetradon cutcutia*	Flat fish.
62.	(Puffer fishes)		

Source: M. Kumari and Yadav, S.C., 2006.

Table 11.5: List of Ornamental Fishes found from the Reefs of Saurashtra, Gujarat

Sl.No.	Family	Species	Months of Availability
1.	Acanthuridae	*Acanthurus santhopterus*	April–June
2.	Ambassidae	*Amnbassis* spp.	Whole year
3.	Apogonidae	*Apogon multitaeneus*	Whole year
4.		*Apogon* spp.	Whole year
5.	Balistidae	*Abalistes stellatus*	Aug.–October
6.		*Odonus niger*	Whole year

Contd...

Table 11.5–Contd...

Sl.No.	Family	Species	Months of Availability
7.	Bothidae	*Pseudorhombus elevatus*	Whole year
8.	Carangidae	*Alectis indicus*	August–November
9.		*Parastromateus niger*	August–December
10.	Chaetodontidae	*Chaetodon collaris*	April–June, November–January
11.		*C. kleinii*	April–June
12.		*C. lunula*	April–June
13.		*Heniochus dispar*	April–June
14.	Cypriniodontidae	*Cyprinidon dispar*	Whole year
15.	Drepanidae	*Drepane punctatus*	Whole year
16.	Echeneidae	*Echeneis naucrates*	Whole year
17.	Gerridae	*Gerres acinaces*	Whole year
18.		*Gerres filamentosus*	Whole year
19.	Gobidae	*Gobiosalbo punvtatus*	Whole year
20.	Haemulidae	*Pomadasys stridens*	Whole year
21.	Labridae	*Heliochoeres dussumieri*	Whole year
22.		*Thalassoma labroids*	Whole year
23.		*T. lunare*	Whole year
24.	Lethrinidae	*Lethrinus rubrioperculatus*	August- October
25.		*L. nebulosus*	August–October
26.	Lutjanidae	*Lutjanus ehrenbergii*	Whole year
27.		*L. monostigma*	Whole year
28.		*L. vitta*	Whole year
29.	Monocanthidae	*Paramonocanthus cingalensis*	Whole year
30.	Mugilidae	*Mugil cephalus*	August–December
31.	Mullidae	*Upeneus vittatus*	Whole year
32.	Pempheridae	*Pempheris mangula*	October–November
33.	Plotosidae	*Plotosus lineatus*	October–November
34.	Pomacanthidae	*Pomocanthus annularis*	Whole year
35.		*P. striatus*	September–January
36.	Pomacentridae	*Abudefduf bengalensis*	Whole year
37.		*A. sexfasciatus*	Whole year
38.		*A. sordidus*	Whole year
39.		*Chrysiptera unimaculata*	December–February
40.		*Neopomacentrus cyanomos*	Whole year
41.		*N. sindensis*	Whole year
42.	Scaridae	*Scarus russelli*	April–August; November–January
43.	Scatophadidae	*Scatophagus argus*	Whole year

Contd...

Table 11.5–Contd...

Sl.No.	Family	Species	Months of Availability
44.	Sciaenidae	*Johnius dussumieri*	Whole year
45.		*J. elongates*	Whole year
46.	Serranidae	*Cephalopholis Formosa*	August–October
47.	Siganidae	*Siganus canaliculatus*	April–August; November–January
48.	Sillaginidae	*Sillago sihama*	Whole year
49.	Terapoidae	*Lagocephalus inermis*	Whole year
	Crustaceans		
50.	Alpheidae	*Alpheus strenus*	Whole year
51.	Palinuridae	*Panulirus homarus*	September-November
52.		*P. polythagus*	September-November
53.		*P. versicolor*	September-November

Source: Joe and Fofandi, 2003.

Table 11.6(a): Native Fish of Jammu having Potential for Ornamental Fish Culture

Sl.No.	Name of the Species
1.	*Bagarius bagarius* (Ham-Buch)
2.	*Barilius bendelisis* (Ham-Buch)
3.	*B. vagra* (Hamilton)
4.	*Botia dayi* (Hamilton)
5.	*Channa orientalis* (Bloch and Schneider)
6.	*Channa punctatus* (Bloch)
7.	*Dani devaria* (Ham-Buch)
8.	*Danio rerio* (Hamilton)
9.	*Esomus dandricus* (Ham-Buch)
10.	*Glossogobius giurls* (Ham-Buch)
11.	*Heteropneustes fossilis* (Block)
12.	*Lepidocephalichthys guntea* (Ham)
13.	*Mastacembelus armatus* (Lacepede)
14.	*Mystus beleekari* (Day)
15.	*M. seenghala* (Sykes)
16.	*M. vittatus* (Bloch)
17.	*Nemacheilus botia* (Hamilton)
18.	*Notopterus chitala* (Ham-Buch)
19.	*Notopterus notopterus* (Pallas)
20.	*Ompak bimaculata* (Bloch)
21.	*Puntius conchonius* (Ham-Buch)
22.	*Puntius sophore* (Ham-Buch)
23.	*Puntius ticto* (Ham-Buch)

Contd...

Table 11.6(a)–Contd...

Sl.No.	Name of the Species
24.	*Rasbora rasbora* (Ham-Buch)
25.	*Tort or* (Ham-Buch)
26.	*Tor putitora* (Ham-Buch)
27.	*Trichogaster fasciatus* (Bloch and Schn)
28.	*Tryploplysa yasinensis* (Alock)
29.	*Xenentodon cancila* (Hamilton)

Source: Charak and Fayaz (2005).

Table 11.6(b): Other Ornamental Fishes at Departmental Aquarium Centre, Jammu

Sl.No.	Name of the Species
Freshwater Fishes	
1.	*Astronotus ocellatus* (Oscar)
2.	*Balantiocheilus melanopterus* (Silver shark)
3.	*Cyprinus* sp. (Koi carp)
4.	*Crassius auratus* (Gold fish)
5.	*Gymnocorymbus ternetzi* (Window etra)
6.	*Helostoma temminck* (Kissing gourami)
7.	*Hemichromis bimaculatus* (Neon jewels)
8.	*Hyphessobrycon* sp. (Tetras)
9.	*Hypostomus multiradiatus* (Sucker catfish)
10.	*Labeo bicolor* (Red tailed black shark)
11.	*Melanochromis auratus* (Auratus)
12.	*Poecilia letipinna* (Mollies)
13.	*Poecilia reticulata* (Guppies)
14.	*Pterophyllus scalare* (Angels)
15.	*Symphysodon aequifasciatia* (Discus)
16.	*Serrasamus nallereri* (Piranha)
17.	*Trichogaster trichopterus* (Blue gourami)
18.	*Trichogaster microlepus* (Moonlight gourami)
19.	*Xiphophorus hatter* (Sword tail)
20.	*Xiphophorus maculatus* (Platy)
21.	*Puntius tetrazona* (Tiger barb)
22.	*Mylosoma plusiventre* (Silver dollar)
Marine Fishes	
1.	*Amphiphrios ocellaris* (Clown fish)
2.	*Chromis cyaneae* (Blue damsel)
3.	*Crysiptera parasema* (Yellow tailed damsel)
4.	*Dascyllus trimaculatus* (Three spot damsel)
5.	*Pterois volitans* (Lion fish)
6.	*Zebrasoma xanthurum* (Sail fish tang)

Source: Charak and Fayaz (2005).

Table 11.7(a): Critically Endangered Freshwater Fishes of Kerala

Sl.No.	Name of the Species
1.	Amblypharyngodon chakaensis
2.	Horabagrus nigricocllaris
3.	Horaglanis krishnaii
4.	Horalabiosa joshuai
5.	Lepidopygopsis types
6.	Silurus wynaadensis
7.	Labeo ariza
8.	Neolissochilus wynaadensis
9.	Ompok malabaricus
10.	Osteochilichthys longidorsalis
11.	Pangasius pangasius
12.	Tor mussullah
13.	Balitora mysorensis
14.	Channa micropeltes
15.	Dayella malabarica
16.	Glyptothorax anamalaiensis
17.	Homalaptera Montana
18.	Puntius ophicephalus

Table 11.7(b): Endangered Freshwater Fishes of Kerala

Sl.No.	Name of the Species
1.	Chela fasciata
2.	Garra hughi
3.	Glyptothorax davissinghi
4.	Gonoproktopterus micropogon periyarensis
5.	Homalaptera menoni
6.	Horadandia attukorali
7.	Ostcochilus thomassi
8.	Ostochilus brevidorsulis
9.	Punrius thomassi
10.	Silaniachildreni
11.	Travancoria clongata
12.	Travancoria jonesi
13.	Top pulitora
14.	Anguilia bengalensis
15.	Esomus thermoicos

Contd...

Table 11.7(b)–Contd...

Sl.No.	Name of the Species
16.	*Garra McClellandi*
17.	*Garra surendranathinii*
18.	*Gonoprokiopterus kolus*
19.	*Gonoproktopterus thomassi*
20.	*Gonoproktopterus kurali*
21.	*Labeo dussumieri*
22.	*Nemacheilus evezardii*
23.	*Nemacheilus monilis*
24.	*Osteobrama bakeri*
25.	*Pangio goensis*
26.	*Puntius lithopidos*
27.	*Puntius melanostigma*
28.	*Sicyopterus griseus*
29.	*Balasio travancoria*
30.	*Glyptothorax annamalacnsis*
31.	*Glyptothorax annandali*
32.	*Gonoproktopterus curmuca*
33.	*Horabagrus brachysoma*
34.	*Puntius denisonti*

Table 11.7(c): Some of the Endemic Freshwater Fish Germplasm Resources of Kerala

Sl.No.	Name of the Species
1.	*Ambassis gymnocephalus*
2.	*Amblypharyngodon chakaensis*
3.	*Amblypharyngodon mola*
4.	*Anabas testudineus*
5.	*Aplocheilus blocki*
6.	*Aplocheilus lineatus*
7.	*Awavous gutum*
8.	*Balitora brucei*
9.	*Balitora mysorensis*
10.	*Barilius barna*
11.	*Batasio travancoria*
12.	*Chanda nama*
13.	*Crossochelius latius latius*
14.	*Dayella malabarica*

Contd...

Table 11.7(c)–Contd...

Sl.No.	Name of the Species
15.	*Eleotris fusca*
16.	*Etroplus maculatus*
17.	*Garra gotyla*
18.	*Garra hughi*
19.	*Garra McClellandi*
20.	*Garra menoni*
21.	*Garra mullya*
22.	*Glyptothorax annandalei*
23.	*Glyptothorax davissinghi*
24.	*Horadandia atukorali*
25.	*Horaglanis krishnai*
26.	*Horalabios joshuai*
27.	*Lepidocephalus thermalis*
28.	*Macropodus cupanus*
29.	*Microphis concalus*
30.	*Nemacheilus evezardii*
31.	*Nemacheilus guentheri*
32.	*Nemacheilus nilgiriensis*
33.	*Osteobrama cotio peninsularis*
34.	*Osteochilichthys longidorsalis*
35.	*Osteocheilus brevidorsalis*
36.	*Pangasius pangasius*
37.	*Pangio baashai*
38.	*Panio goensis*
39.	*Parambassis dayi*
40.	*Pseudambassis ranga*
41.	*Puntius amphibius*
42.	*Puntius dorsalis*
43.	*Puntius melanostigma*
44.	*Puntius pinnuratus*
45.	*Salmostoma boopis*
46.	*Salmostoma clupeoides*
47.	*Amblypharyngodon melettinus*
48.	*Amblypharyngodon microlepis*
49.	*Barilius bakeri*
50.	*Barilius bendelesis*
51.	*Barilius canarensis*

Contd...

Table 11.7(c)–Contd...

Sl.No.	Name of the Species
52.	*Barilius gatensis*
53.	*Bhavania australis*
54.	*Chela dadiburjori*
55.	*Chela fasciata*
56.	*Chela laubuca*
57.	*Danio aequipinnatus*
58.	*Danio malabaricus*
59.	*Esomus danricus*
60.	*Esomus thermoicos*
61.	*Glyptothorax housei*
62.	*Glyptothorax lonah*
63.	*Homalaptera menoni*
64.	*Homalaptera montana*
65.	*Homalaptera pillai*
66.	*Lepidopygopsis typus*
67.	*Mystus oculatus*
68.	*Nandus nandus*
69.	*Nemacheilus menoni*
70.	*Nemacheilus pulchellus*
71.	*Nemacheilus striatus*
72.	*Nemacheilus petrubenarescui*
73.	*Parambassis thomassi*
74.	*Pristolepis fasciata*
75.	*Pristolepis marginata*
76.	*Puntius chola*
77.	*Puntius fasciatus*
78.	*Puntius filamentosus*
79.	*Puntius lithopidos*
80.	*Puntius singhala*
81.	*Puntius sophore*
82.	*Puntius ticto*
83.	*Puntius vittatus*
84.	*Rasbora daniconius*
85.	*Salaries reticulates*
86.	*Salmostoma acinaces*
87.	*Sicyopterus griseus*
88.	*Tettradon travancoricus*

Contd...

Table 11.7(c)–Contd...

Sl.No.	Name of the Species
89.	*Travancoria jonesi*
90.	*Travancoria elongata*
91.	*Garra surendranathinii*
92.	*Glyptothorax anamalaiensis*
93.	*Nemacheilus botia*
94.	*Nemacheilus denisoni denisonii*
95.	*Nemacheilus keralensis*
96.	*Nemacheilus monilis*
97.	*Nemacheilus pambarensis*
98.	*Nemacheilus periyarensis*
99.	*Nemacheilus semiarmatus*
100.	*Nemacheilus triangularis*
101.	*Osteobrama bakeri*
102.	*Puntius chalakkudiensis*
103.	*Puntius conchonius*
104.	*Puntius denisonii*
105.	*Puntius jerdoni*

Source: Madhusoodan, Krupa B. (2003).

Table 11.8: Ornamental Fishes of Rajasthan

There are 147 species of freshwater fishes reported from Rajasthan. Main genus of the same is as under.

Sl.No.	Name of Genus
1.	*Danio*
2.	*Esomus*
3.	*Oxygaster*
4.	*Chella*
5.	*Rasbora*
6.	*Puntius*
7.	*Trichogaster*
8.	*Amblypharyngodon*
9.	*Namachelius*
10.	*Lepidocephalichthys*
11.	*Labio*
12.	*Cirrhinus*
13.	*Mastocembalus*

Contd...

Table 11.8–Contd...

Sl.No.	Name of Genus
14.	*Channa*
15.	*Notopterus*
16.	*Ompok*
17.	*Mystus*
18.	*Xenanthodon*
19.	*Cyprinus*
20.	*Glossgobius*
21.	*Guirus*
22.	*Tilapia*
23.	*Garra*
24.	*Therapon*
25.	*Cichilid*
26.	*Etroplus*

Source: Durva, V.S., 2005.

Table 11.9: Exportable Ornamental Fishes Identified in West Bengal

Sl.No	Local Common Name	Scientific Name
1.	Spotted moray eel	*Lycodontis tile*
2.	Devil catfish	*Chhaca chaca*
3.	Topaz pusser	*Chelendon spelenodachneri*
4.	Red green dwarf puffer	*Monotrentus travancoricus*
5.	Juguar loach	*Somileptus gongota*
6.	Spiny green eel	*Mastacembelus pancalus*
7.	Dwarf gourami	*Coisa fasciatus*
8.	Jewel glass fish	*Chanda ranga*
9.	Mourala	*Amblypharyngodon mola*
10.	Techokho	*Aplocheilus panchax*
11.	Chang	*Channa gachua*
12.	Lata	*Channa punctatus*
13.	Nandas	*Nandus nandus*
14.	Panther loach	*Lepidocephalichthys guntea*
15.	Leopard loach	*Acanthocobitis botia*
16.	Spotted loach	*Noemachelius corcia*
17.	Banded loach	*N. savona*
18.	Zebra loach	*N. zonaius*

Source: Radha C. Das and Archna Sinha, 2003.

Appendix I

Contd...

Appendix I–Contd...

Sl.No.	Family	Common Name	Scientific Name	Figure No.	Page No.
16		Burmese glass fish	*C. baculis*	7.E.3	100
17		Elongate glass perch	*C.nama*	7.E.4	101
F	Channidae	SNAKEHEADS			
18		Rainbow snakehead	*Channa bleheri*	7.F.1	102
19		Burmese snakeghead	*C. burmanica*		
20		Forest nakehead	*C. lucius*	7.F.2	103
21		Dwarf snakehead	*C. gachua*	7.F.3	103
22		Giant snakehead	*C. micropelets*	7.F.4	104
23		Smooth breasted snakehead	*C. orientalis*	7.F.5	104
G	Characidae	GOURAMI			
24		Pearl gourami	*Trichogaster leeri*	7.G.1	105
25		Kissing gourami	*Helostoma termmincki*	7.G.2	107
26		Dwarf gourami	*Colisa lalia*	7.G.3	108
27		Three spot gourami	*Trichogaster trichopterus*	7.G.4	109
28		Blue gourami		7.G.5	110
29		Chocolate gourami	*Sphearichthus osphoromoenoides*	7.G.6	110
30		Honey gourami	*Colisa chuna*	7.G.7	112
31		Indian gourami	*C. fasciata*	7.G.8	113
32		Thicked lipped gourami	*C. labiosa*		
G	Characidae	HATCHET FISH			
33		Marbled hatchet fish	*Carnegiella strigata*	7.G.9	114
34		Silver hatchet fish	*Gasterolepecus strernicta*	7.G.10	116
G	Characidae	TETRA			
35		Cardinal tetra	*Cheirodon exelrodia*		
36		Black window tetra	*Gymnocorymbus serape*		
37		Serpae tetra	*Hyphessobrycon serape*		
38		Head & tail light tetra	*Hemigrammus ocellifer*		
39		Red eye tetra	*Moenkhaqusia sanctaefilomenae*		
40		Neon tetra	*Paracheirodon innesi*	7.G.11	120
41		Hocky stick tetra	*Thayeria bochlkei*		
42		Black neon tetra	*Hyphessobrycon herbertaxelrodi*		
43		Bleeding tetra	*Hypessobrycon erythrostigona*		

Contd...

Appendix I–Contd...

Sl.No.	Family	Common Name	Scientific Name	Figure No.	Page No.
44		Blood fin tetra	*Aphyocharax anisitsi*		
45		Red head tetra	*Petitella georgiae*		
46		Lemon tetra	*Hypessobrycon pulchripinnis*		
H	Cichlidae	CICHILIDS			
47		Chilid	*Pseudotrophenus lambarddoi* (male & Female)	7.H.1.1 7.H.1.2	126
126					
48		Nyasa golden cichlid	*Melanochromis aurathus*	7.H.2	127
49		Blue acarea	*Aegruidens latiforms*		
50		Angel	*Pterophyllum scalar* (22 varieties)	7.H.3.1 to 7.H.3.14	130 to 132
51		Flower horn	*Cichlasoma cichlids* (Hybrid)		
52		*Orientental beauty*			
53		Galaxy Blue			
54		Happy star	*Cichlasoma Spp.*	7.H.4	135
55		Storm Rider	*Cichlasoma Spp.*	7.H.5	135
56		Wonder spark	*Cichlasoma Spp.*	7.H.6	135
57			*Lamprologus stappersi*		
58		Royal tiger			
59		Red beauty			
60		Moon light beauty			
61		The may blossom			
62			*Julidochromis ornatus*		
63			*Tropheus duboisi*		
H	Cichlidae	DISCUSS			
64		Pigeon blood		7.H.7	140
65		Blue turquoise	*S. acquifaciata hedri*	7.H.8	140
66		Red turquoise		7.H.9	140
67		Brown discuss	*S. acquifaciata axeirodi*	7.H.10	140
I	Cypriniformes	BOTIA / LOACH			
68		Clown loach	*Botia macrocanthius*	7I.1	142
69		Dwarf loach	*B. sidthimunk*	7.I.2	143
70		Yo yo loach	*B. almorhae*	7.I.3	143
71		Golden zebra loach	*B. histrionica*	7.I.4	144
72		Tiger loach	*B. hymenophysa*	7.I.5	145
73		Queen loach	*B. dario*	7.I.6	145
74		Polka dotted loach	*B. beauforti*	7.I.7	145

Contd...

Appendix I–Contd...

Sl.No.	Family	Common Name	Scientific Name	Figure No.	Page No.
JB	Cyprinidae	BARBS			
75		Tiger barb	*Barbus tetrazona*	7.J.1	147
76		Rosy barb	*B. conchanium*	7.J.2	147
77		Flying barb	*Esomus danrica*	7.J.3	148
78		Black spot barb	*Putius filamentosus*	7.J.4	149
79		Long fin barb	*Capaeta arullas*	7.J.5	150
80		Tic – tac too barb	*Puntius ticto*	7.J.6	151
81		Black ruby barb	*P. titteye*	7.J.7	151
82		Green barb		7.J.8	151
83		Tin fin barb		7.J.9	152
JD	Cyprinidae	DANIO			
84		Giant danio	*Danio malbaricus*	7.JD.1	153
85		Pearl danio	*D. albolineata*	7.JD.2	154
86		Spotted danio		7.JD.3	155
87		Zebra danio	*Brachydanio rerio*	7.J.D.4 (a & b)	156
JD	Cyprinidae	GOLD FISH			
88		Shubunkin		7.JG.1 & 7.JD1.1	159
89		Comet		7.JG.2	160
90		Fantail		7.JG.3	160
91		Viltail fin Gold		7.JG.4.	160
92		Pearl scale		7.JG.5	161
93		Black moor		7.JG.6	161
94		Celestial		7.JG.7	161
95		Oranda		7.JG.8	162
96		Lion head		7.JG.9	162
97		Bubble eye		7.JG.10	163
98		Jikin		7.JG.11	163
99		Ryukin			
100		Tosakin		7.JG.12	164
		Varieties of Gold Fish		6 Nos	165 & 166
101		Koi carp	*Cyprinus carpio*		172
JR		RAINBOW FISH			
102		Jewel Rainbow	*Malantaenia trifasciata*	7.JR.1	174
103		Lake Kutubu Rainbow	*M. lacustris*		

Contd...

Appendix I–Contd...

Sl.No.	Family	Common Name	Scientific Name	Figure No.	Page No.
104		Arfak Rainbow	*M. arfakensis*	7.JR.2	174
105		Boeseman's Rainbow	*M. boesemani*		
JR	Cyprinidae	RASBORA			
106		Pygmy rasbora	*Rasbora maculate*	7.JR.3	175
107		Red tailed rasbora	*R. borapetensis*		
108		Hari-quin rasbora	*R. hetreromorpha*	7.JR.4	175
109			*R. pauiperforata*	7.JR.5	176
110			*R. trilineata*	7.JR.6	176
111			*R. partluciosoma libosa*	7.JR.7	176
K	*Cyrinidontif*	PANCHAX			
112		Stripped panchax	*Aplochelius .A. lineatus*	7.K.1	178
113		Rainbow panchax	*A. blockii*	7.K.2	179
114		Blue panchax	*A. panchax*	7.K.3	179
115		Green panchax	*Apolochelius dayi*	7.K.4	179
L	Mestacem-belidae	EEL			
116		Spiny eel	*Macrognathus aculeathus*	7.L.1	181
117		Zigzag eel	*Mastracembelus armatus*	7.L.2	182
M	Nandidae	BADI			
118			*Badies badies badies*	7.M.1	183
119			*B. badies salnensis*	7.M.2	183
120			*B. badies bunanicus*	7.M.3	184
N	Poiciliidae	MOLLY			
121		Black molly	*Poicillia sphenops*	7.N.1	185
122		White Molly	*Poicillia sphenops*		
123		Marbal molly	*Poicilia latipinna*	7.N.2	187
124		Golden Ballon molly	*Xiphophottus Variatus*	7.N.3	187
125		Sward tail	*Xiphophorus helleri*	7.N.4	188
126		Platy	*Xiphophorus maculates*	7.N.5	190
127		Spotted molly	*P. latipinna*		
O		GUPPY			
128		Million fish	*Poicilia reticulatat*	7.O.1	191
129		Guppy male			
130		Black Guppy			193
131		Delicate Variegated Guppy			193
132		Green Variegated Guppy			193

Contd...

Appendix I–Contd...

Sl.No.	Family	Common Name	Scientific Name	Figure No.	Page No.
133		Blue Variegated Guppy			193
134		Red Snakeskin Guppy			194
135		Green Dankeskin Guppy			194
136		Yellow Snakeskin Guppy			194
137		Tuxedo Guppy			195
138		Tuxedo Cobra Guppy			195
139		Red Tailed Guppy			195
140		Blond Red Tail Guppy			196
141		Blue Tail Guppy			196
142		Pineapple Guppy			196
143		Double Sward Guppy			197
144		Electric Blue Guppy			197
145		Leopard Guppy			197
146		Lyretail Multicolor Guppy			198
147		Mosquito fish	*Gambusia affinis affinis*	7.O.2	199
P		OSCAR			
148		Oscar	*Astronotus ocellatus*	7.P.1	201
Q	Tetraodon-tidae	PUFFER FISH			
149		Puffer fish	*Monotetrus travan*	7.Q.1	203
150		Blow fish	*Tetradon fluviatilus*	7.Q.2	203
151		Ballon fish	*T. nigroviridis*	7.Q.3	204
152			*T. mbu*	7.Q.4	204
153			*T. biocellatus*		
154			*T. leiurus*		
155			*T.lineatus*		
156			*Cariotetraodon . somphongs*		
157	NA	Mouth Brooder	*Tilapia Mosambica*	7.R.1	206
158	NA	Red Tail Black shark	*Labio bicolor*	7.S.1	207

Over and above Photographs of Angel (14) and 6 varieties of Gold fish are given at respective places. Hence total fish covered are 178.

Details of Figure number:

7: Freshwater Fish

A to S: Family No.

1,2,3,: Fish number in respective family.

Appendix II

Contd...

Appendix II–Contd...

Sl.No.	Family	Common Name	Scientific Name	Figure No.	Page No.
F	Grammidae				
14		Royal Gramma	Gramma loreto		
15		Black cap Gramma	G. melacare		
G	Labridae				
16		Wrasses	Cirrhilabrus ruhriventralis		
17		Pyjama wrasses	Pseudocheilinus hexataenia	8.12	225
H	Opistognthidae				
18			Opistognathus auriforms	8.13	226
I	Pomacentridae				
19			Amphirion clerkii	8.14	227
20		Green chormis	Chormis caerules	8.15	228
J	Pomacanthiidae				
21		Oriode Angel	Centropyge bicolor	8.16	229
22		Flame Angel fish	C. loriculus	8.17	230
23		Blue Ring Angel fish	Pomacanthus annularis	8.18	231
24			P. semicirculatus	8.19	231
K	Syngnathidae				
25		Sea Horse	Hippocanpus erectus	8.20	232
26		Sea Horse	H huda	8.21	232
L	Opistognthidae				
27		Red Arowana	Opistognathus auriforms	8.22	233
28		Golden Arrowana		8.23	234
29		Red tailed Gold Dragon			
30		Green Dragon			
M	Sluridae				
31		Stripped Cat Fish	Mystus vittatus	8.24	235

Glossary

Important Words Used in Aquaculture

Sl.No	Word	Meaning
1.	Acclimatization	Individual adaptation of living things to changed conditions of life, especially to change in climate or the physico-chemical parameters of their environment.
2.	Abiotic	Nonliving components of environment or ecosystem which invariably includes physico-chemical factors such as light, temperature, dissolved gases and nutrients (micro and macro).
3.	Adipose fin	Small, fleshy fin found between the dorsal fin and the tail fin in some fishes like Piranhas.
4.	Aerobic	Requiring oxygen for survival (respiration) and growth.
5.	Algae	A group of green plants without roots stem or leaves. Some are single celled but others are many celled.
6.	Anaerobe	Bacteria which can live and multiply without oxygen.
7.	Anal fin	An unpaired fin between the anus and the tail, on the underside of the fish.
8.	Aquarium	An aquarium is a glass – tank with standing water in which aquatic animals and / or plants are kept generally for exhibition.
9.	Arobibenthic	Refers to forms inhabiting the sea bottom below the edge of the continental shelf.

Sl.No	Word	Meaning
10.	Autotropic	Producing its own food; photosynthetic plants but also certain bacteria in aquatic system.
11.	Barbel	Fishy beard – like organ found attached to the mouth region of some fishes used for detecting food by tests.
12.	Benthos	Organisms living on the bottom of water body. Benthos of plant origin are phyto benthos while those of animal origin are referred as zoo benthos.
13.	Biomass	The wet or dry weight of living organisms.
14.	Biota	All the living elements of an ecosystem or given area.
15.	Brackish water	Water containing approximately 10 percent sea water; found in estuaries where freshwater rivers enter the sea.
16.	Biological filtration	Means of water filtration using bacteria, *Nitrosomonas* and Nitrobacteria to reduce otherwise toxic ammonium based compounds to safer substances such as nitrates.
17.	Caudal Peduncle	Part of fish's body joining the caudal fin to the main body.
18.	Debris	Rubbish, garbage, anything that is not supposed to be in a certain area.
19.	Diatom	Member of class *Bacillariophyceae* of algae that possesses a wall of overlapping silica valves.
20.	Detritus	Decaying organic matter found on the bottom of the pond / tank .
21.	Dropsy	Accumulation of fluids in the abdomen; edema.
22.	Euphotic zone	The upper layers of water in which the light is sufficient for photosynthesis.
23.	Ectoparasite	Gill parasites are usually referred as ectoparasite.
24.	Endoparasite	They live inside the body of the host.
25.	Filter feeder	Animal that sifts water for microscopic food.
26.	Fin rot	Bacterial ailment, the tissue between the rays of the fin rots away.
27.	Genital papillae	They are breeding tubes that extend from the vet of each fish; usually larger I females than males.
28.	Gonopodium	Anal fin that has be modified to form a copulatory organ male live bearing fishes.
29.	Hybrid	A cross-bred offspring, coming from parents of different hereditary characteristics; generally hybrids show greater vigor; but are sterile in some cases.

Sl.No	Word	Meaning
31.	Hypophysation	The process of inducing the fish for breeding, by injecting pituitary extract or sex hormones.
32.	Iris	The rim of the eye in fishes.
33.	Larvophiles	They are larvae loving mouth brooders lay their eggs on a substrate ad guard them until the eggs hatch.
34.	Lateral line	Line of perforated scales along the flanks which lead to a pressure sensitive nervous system.
35.	Lithophillic eggs	Eggs often sticking to rocks or other hard substratum
36.	Live bearers	The fish or animal that give birth to live young ones.
37.	Littoral zone	The shallow zone ear shore of a body of water; out to the usual limit of influences of wave action or tides, with daylight reaching the bottom life.
38.	Mucous	Slimy secretion by membranes to moisten and protect the skin.
39.	Nauplii	Typical larvae of crustacean formed directly from the egg.
40.	Nocturnal	Animals active during night time.
41.	Operculum	The bony gill cover of fishes.
42.	Ovophies	They are egg loving mouth brooders; lay their eggs in a pit.
43.	Oviparous	Animals that lay eggs.
44.	Ovoviviparous	The eggs hatched within the female before they are released.
45.	Pathogen	Organism or anything capable of producing diseases.
46.	Paludarium	A paludarium is a shore aquarium which is not all water but has some dry land rising out to simulate a shore.
47.	Peat	An acid soil derived from decomposed plants, especially from moss.
48.	Pectoral fins	The paired fins placed behind the gill openings of fishes.
49.	Pelvic fin	The paired fins attached to the under side of the fish, anterior to the anus.
50.	Photophobic	Allergy to sunlight.
51.	ppm	Parts per million, a measurement of concentration.
52.	ppt	Parts per thousand a measurement of concentration.

Sl.No	Word	Meaning
53.	Plankton	Small organisms occupying the top layers of water with feeble power of movement; often transported by water currents.
54.	Prophylaxis	Prevention of disease.
55.	Spawning	The deposition of eggs by aquatic animals.
56.	Species	A basic taxonomic unit of animals and plants.
57.	Symbiosis	Relationship between two parties each deriving mutual and indispensable benefit. Advanced form of commensalisms.
58.	Terrarium	It is vivarium without standing water.
59.	Thermophile	Organism tolerant of high temperature of hot spring and streams. Organism that live in hot springs and streams.
60.	Tubifex worms	A group of small fresh water segmented worms that live on mud or other decaying matter.
61.	Ulcer	Sore on skin or internal organ of the body.
62.	Ventral fin	Paired fins on the underside of a fish in front of the anus.
63.	Vivarium	A vivarium is an enclosure or container for keeping, observing and/or displaying living animals and plants.
64.	Zooplankton	Minute occupying the upper layers of the water column.

References

Anon. 1998. Training Manual on Culture of live food organisms for Aqua hatcheries. Central Institute of Agricultural Research Education, Mumbai.

Anon. 1999. How to develop ornamental Fish farming. Seafood Export Journal Vol 30 (2): 31.

Alagappan Mand Vigula K. 2004. Aquarium Fish Breeding Techniques. Fishing Chimes, 24 (5): 26–27.

Amita sexena. 2003. Aquarium Management. Book published by M/s Daya Publishing House, New Delhi.

Atul Kumar Jain. 2005. Aquarium Hardware Course Manual–winter school by ICAR 8–25 Feb: 49–56.

Barry James. 2000. Aquarium Plants. Book published by Interpet Ltd, U. K.

Bodre Ravindra B. and L.L. Sharma. 2005. Breeding and culture of Angel Fish *Pterophyllum salare*. Course Manual–winter school by ICAR 8–25 Feb: 1–5.

Charak K. S. and F. A. Fayaz. 2005. Ornamental fishes of Jammu. Fishing Chimes 25 (6): 24–25.

Chairs Andrew. 1986. Guide to Fish Breeding. Book published by Interpet Ltd., U. K.

Dholakia, A. D. 2000. Aquarium fish and their maintenance. Matsya Gandha published by College of Fisheries, Gujarat Agricultural University Vol III: 20–21.

Dholakia A. D. and A. Y Desai. 2004. Aquarium and its maintenance (in Gujarati). Folder published by College of Fisheries, Junagadh Agricultural University, Veraval.

Dholakia A. D. 2005. Economical strengthening with the help of of Aquarium fish breeding. Radio Talk at All India Radio, Rajkot, March 2005.

Dholakia A. D., Brajendra Kumar, Sagar Chandra Mandal, V. P. Saini and Anil Kumar. 2005. Garden Pool. Advances in Culture and Breeding of Freshwater Ornamental fish and Aquarium Management at Udaipur: 9–11.

Durve V. S. 2005. An over view of ornamental fish trade and possibility of its development in Rajasthan. Course Manual–Winter School by ICAR 8–28 Feb. 2:1–6.

Gosh Indrani. 2006. Captive Breeding of Ornamental Fishes. Fishing Chimes 26 (3): 23–28.

Gupta, A. K. 2005. Role of micro nutrients (Vit, and minerals) in production of ornamental fishes and their management. Course Manual–Winter School by ICAR 8–28 Feb: 62–66.

Harishanker, J. Alappat and A. Biju Kumar. 1997. Aquarium Fishes (A colorful profile) Book published by B. R. Publishing Corporation Delhi–52.

Ivan Petrocicky. 1988. Aquarium Fish of the World. Book published by Arch Cape Press, New York.

Infofish International 2/2001: 9 Gold fish.

Do 3/2001: 14–15 Ornamental fish trade overview.

Do 5/2001: Discuss.

Do 1/2002: 18 Koi Carp.

Do 2/2002: 35 Gourami.

Do 3/2002: 21 Barbs.

Do 6/2002 : 22 Arowana.

Do 2/2003 : 15 Rainbow.

Do 3/2003 : 32 Loaches.

Infofish International 4/2003 : 17 Corydoras.

Do 5/2003 : 18 Rasboras.

Keshavanath and Prakash Patel 2006. Nutrition in Ornamental fishes. Fishing Chimes, 26 (8): 13–18.

Kuldip Kumar 2004. Maintenance and Management of fish aquarium. Fishing Chimes, 24 (4): 46–48.

Kumari, M. and Yadav, S. C. 2006. Ornamental fishes of Bihar Wet lands and their export potential. Fishing Chimes, 25 (10): 43.

Mahapatra, B. K., K. Vinod and B. K. Mandal 2005. Indigenous Ornamental inland Fish Resources of North Eastern India. Fishing Chimes 25 (8): 19–24.

Manoj Das and Apurba Kumar Das 2005. Status of Ornamental fishes of Assam. Fishing Chimes 25 (3): 13–17.

Madhusoodan, Krupa B. 2003. Popularization of Aquarium keeping and promotion of ornamental fish culture and marketing: Kerala Model of Development. Fishing Chimes, 23 (2) 31–34.

Nick Dakin 1992. The Book of Marine Aquarium. Book published by Salamander Book Ltd., U. K.

Omprakash Sharma 2005. Food and feeding of Aquarium Fishes. Course Manual–Winter School by ICAR, 8–28 Feb: 86–88.

Peter, W. Scott. 1987. A Fish keeper's guide to live bearing fishes. Book published by Salmander Book Ltd., London, U. K.

Radha, C. Das and Archana Sinha, 2003. Ornamental fish Trade in India (West Bengal and Tripura). Fishing Chimes, 23 (2): 16–18.

Santhanam, R. N. Sukumaran, and P. Natrajan. 1999. A manual of Freshwater aqua culture. Book published by Oxford and IBH Publishing Co. Pvt. Ltd, New Delhi.

Saroj, K. Swain and Partha Bondopadhyay, 2002. Breeding Technology in Ornamental Fish. Fishing Chimes 22(3): 56–60.

S. K. Swain, J. K. Jena and S. Ayyappan. 2000. Prospects of Freshwater Ornamental Fish culture in India with reference to export marketing. Fishing Chimes 20 (10 and 11): 99–101.

Sharma, L. L., S. K. Sharma, B, K. Sharma and V. P. Saini 2005. Tropical Freshwater aquarium fishes. Course Manual–Winter School by ICAR 8–28 Feb: 4–7.

Sharma, L. L. 2005. Aquarium Plants and their culture. Course Manual–Winter School by ICAR 8–28 Feb: 14–16.

Srivastava, C. B. L. 2002. Aquarium Fish Keeping. Book published by Kitab Mahal Allahabad.

Varma, S. K. 2005. Freshwater ecosystem and practices in the ornamental fish aquarium management. Course Manual–Winter School by ICAR, 8–28 Feb: 57–60.

Arrangements in Aquarium (Page 17)

Figure 2.4: Arranging Corals

Figure 2.5: Arranging Plants

Figure 2.6: Arranging Gravels

Different Shapes of Aquarium (Page 18)

Figure 2.9: Normal Shape with Cover

Figure 2.7: Round Shape

Figure 2.8: Octagonal

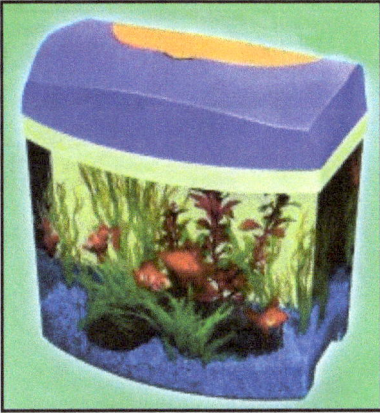

Figure 2.12: Special Shape
(Page 19)

Figure 2.14 (Page 20)

Figure 2.16: Different Types of
Gravels (Page 21)

Figure 3.1(b): Under Gravel Filter
(Actual) (Page 25)

Figure 3.3(b): Sponge Filter
(Actual) (Page 26)

Figure 3.6(a): Canister Power Air Lift Type (Actual) (Page 28)

Figure 3.7: Power Head Filter (Page 29)

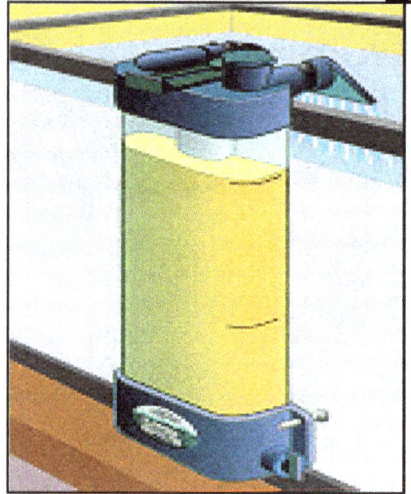

Figure 3.8(a): Fluidized Bed Filters (Page 30)

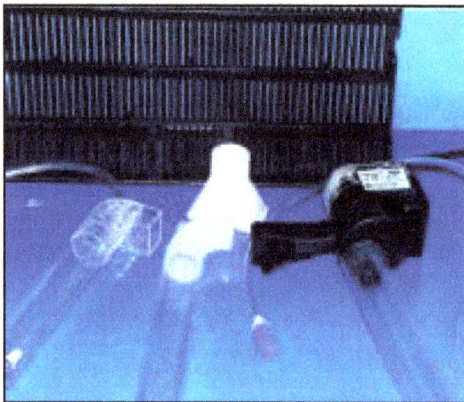

Figure 3.8(b): Under-gravel Filters (UGF) (Page 30)

Figure 3.8(c): Box Filter (Simple variety) (Page 31)

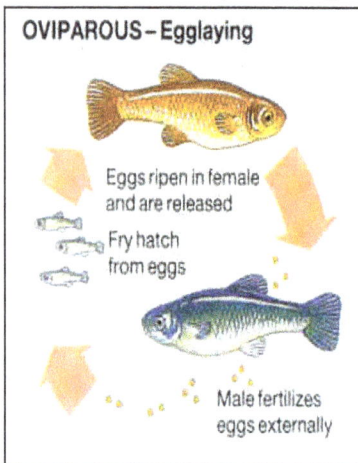

OVIPAROUS – Egglaying

Eggs ripen in female and are released

Fry hatch from eggs

Male fertilizes eggs externally

Figure 6.1 (Page 75)

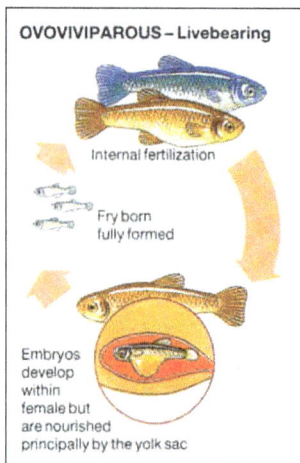

OVOVIVIPAROUS – Livebearing

Internal fertilization

Fry born fully formed

Embryos develop within female but are nourished principally by the yolk sac

Figure 6.2 (Page 75)

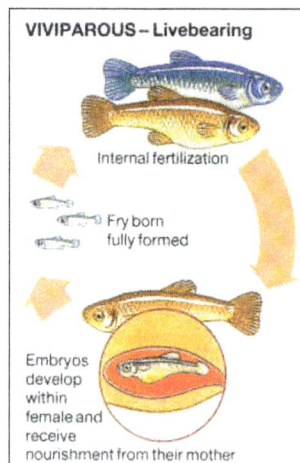

VIVIPAROUS – Livebearing

Internal fertilization

Fry born fully formed

Embryos develop within female and receive nourishment from their mother

Figure 6.3 (Page 75)

Figure 6.8: Mouth Brooder (Page 79)

Figure 6.9: Male Fighter Fish taking Care of their Eggs (Page 80)

Figure 6.11: Male of Fighter Fish Squeezes the Female until She Releases the Eggs (Page 81)

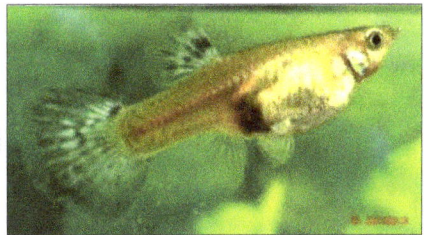

Figure 6.15: Female Guppy Showing Gravid Spot (Livebearer) (Page 83)

Freshwater Ornamental Fishes

Figure 7.A.1: Fighter Fish (Male) (Page 89)

Figure 7.A.1: Fighter Fish (Female) (Page 89)

Figure 7.D.1: *Corydoras schwartzi* (Page 91)

Figure 7.D.2: *Corydoras ambiacus* (Page 91)

Figure 7.D.3: *Corydoras agassizii* (Page 91)

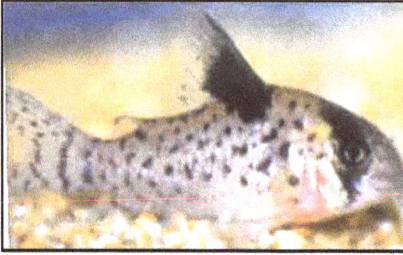
Figure 7.D.4: *Corydoras leucomelas* (Page 91)

Figure 7.E.1: *Chanda ranga* (Page 99)

Figure 7.E.2: *Chanda lala* (Page 100)

Figure 7.E.3: Painted Glass Fish (Page 100)

Figure 7.E.4: *C. Baculis* (Page 101)

Figure 7.F.1: Rainbow Snakehead *Channa bleheri* (Page 102)

Figure 7.F.2: Snakehead *Channa lucius* (Page 103)

Figure 7.F.3: Snakehead *Channa gachua* (Page 103)

6

Figure 7.F.4: Snakehead
Channa micropelles (Page 104)

Figure 7.F.5: Snakehead
Channa orintalis (Page 104)

Figure 7.G.2: Kissing Gourami (Page 107)

Figure 7.G.1: Pearl Gourami (Page 105)

Figure 7.G.3: Red Dwarf Gourami (Page 108)

Figure 7.G.4: Three Spot Gourami (Page 109)

Figure 7.G.6: Chocolate Gourami (*Sphaerochthys osphoromoenoides*) (Page 110)

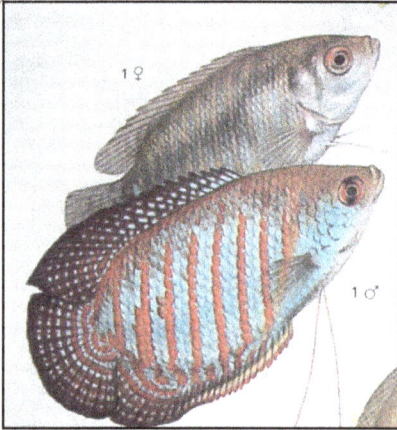

Figure 7.G.5: Blue Gourami (M&F) (Page 110)

Figure 7.G.7: Honey Gourami (Page 112)

Figure 7.G.8: Indian Gourami (Page 113)

Figure 7.G.9: *Carnegiella strigata strigata* (Page 114)

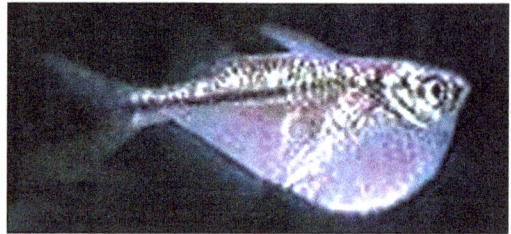

Figure 7.G.10: *Gasterolepecus sternicla* (Page 116)

Figure 7.H.1.1: *Pseudotropheus lombardoi* (Male) (Page 126)

Figure 7.H.1.2: *Pseudotropheus lombardoi* (Female) (Page 126)

Figure 7.H.2: *Melanochrimis auratus* (M and F) (Page 127)

Figure 7.H.3.1: White Angel (Page 130)

Figure 7.H.3.2: Blushing Angel (Page 130)

Figure 7.H.3.3: Koie Angel (Page 130)

Figure 7.H.3.4: Zebra Lace Angel (Page 130)

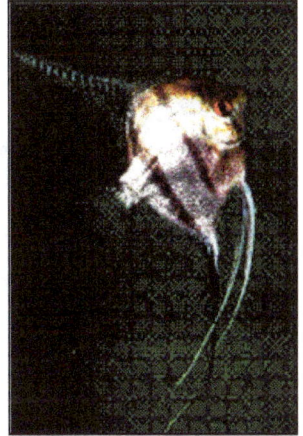

Figure 7.H.3.5: Silver Angel (Page 130)

Figure 7.H.3.6: Koi Angel (Page 130)

Figure 7.H.3.7: Golden Marbel Angel (Page 130)

Figure 7.H.3.8: Yellow Head Angel (Page 130)

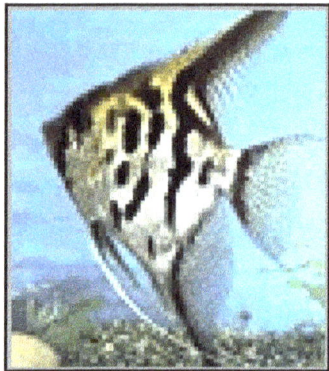

Figure 7.H.3.9: Zebra Angel (Page 131)

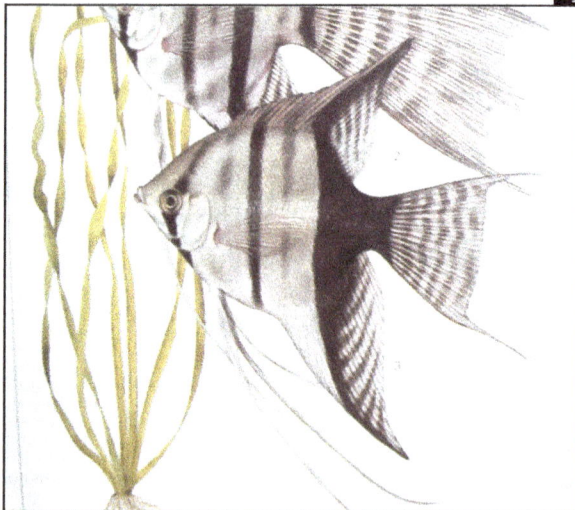

Figure 7.H.3.10: White Marble Angelfish (Page 131)

Figure 7.H.3.11: Black Angelfish (Page 131)

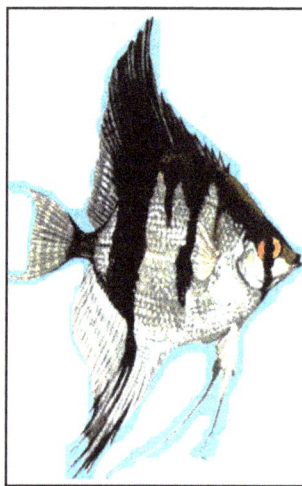

Figure 7.H.3.12: Angel (Page 131)

Figure 7.H.3.13: Angel Fish (Page 132)

Figure 7.H.3.14: Golden Angel Fish (Page 132)

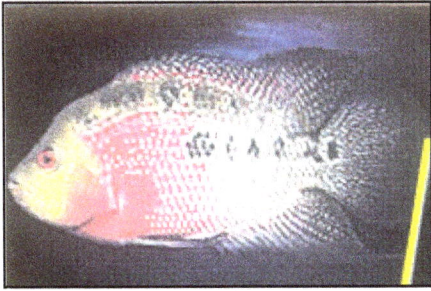

Figure 7.H.4: *Cichlasoma* spp. "Happy star" (Page 135)

Figure 7.H.5: *Cichlasoma* spp. "Strom Rider" (Page 135)

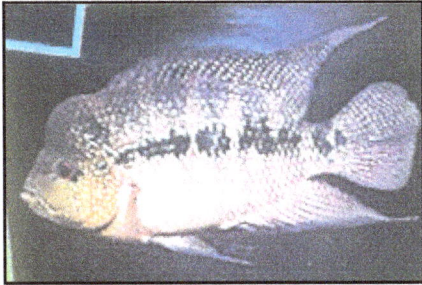

Figure 7.H.6: *Cichlasoma* spp. "Wonder Spark" (Page 135)

Figure 7.H.7: (Discuss) Pigeon Blood (Page 140)

Figure 7.H.9: Red Turquoise (Page 140)

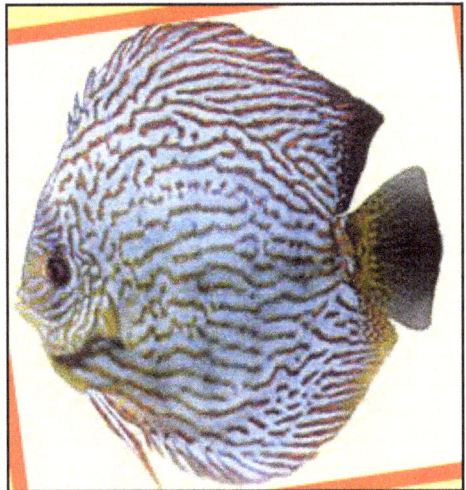

Figure 7.H.8: (Discuss) Blue Turquise (Page 140)

Figure 7.H.10: Brown Discuss (Page 140)

Figure 7.I.1: Clown Loach
(*Botia macrocanthus*) (Page 142)

Figure 7.I.2: Dwarf Loach
(*Botia sidthimunki*) (Page 143)

Figure 7.I.3: Yo Yo Loach
(*Botia almorhae*) (Page 143)

Figure 7.I.5: Queen Loach
(*B. Dario*) (Page 145)

Figure 7.I.4: Golden Zebra Loach (*Botia histrionica*) (Page 144)

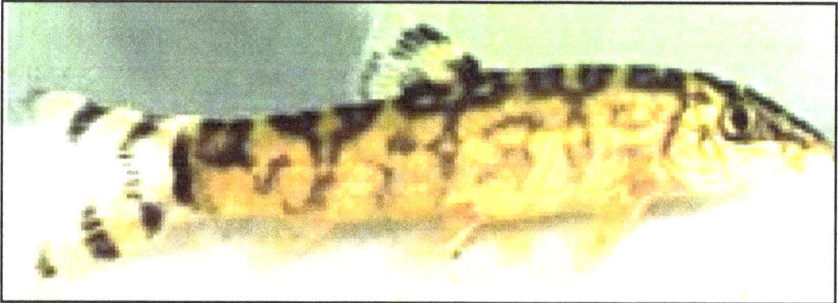
Figure 7.I.6: Tiger Loach (*B. hymenophysa*) (Page 145)

Figure 7.I.7: Polka Dotted Loach (*B. beauforti*) (Page 145)

Figure 7.J.1: Tiger Barb
(*Barbus tetrazona*) (Page 147)

Figure 7.J.2: Rosy Barb
(*Barbus conchonium*) (Page 147)

Figure 7.J.3: Flying Barb (*Esomus danrica*) (Page 148)

Figure 7.J.4: Black Spot Barb (Page 149)

Figure 7.J.5: *Capoeata arullas*
(Male upper) (Female lower) (Page 150)

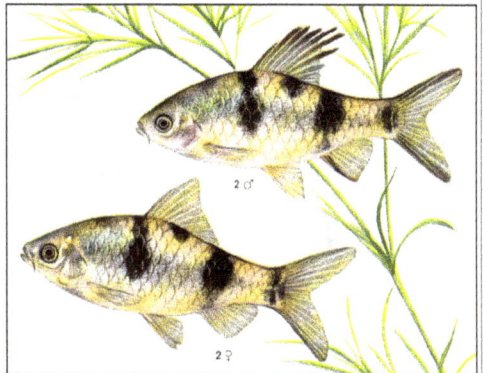

Other Species of Bark

Figure 7.J.6: Tic-TacToo Barb
(*Puntius ticto*) (Page 151)

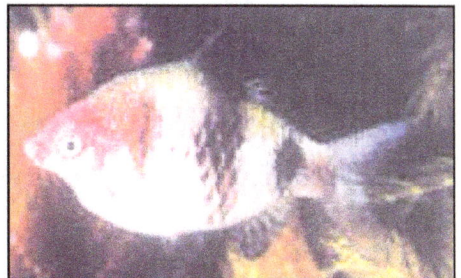
Figure 7.J.7: Black Ruby Barb
(*Punctius titteye*) (Page 151)

Figure 7.J.8: Green Tiger Barb (Page 151)

Figure 7.J.9: Tinfin Barb (Page 152)

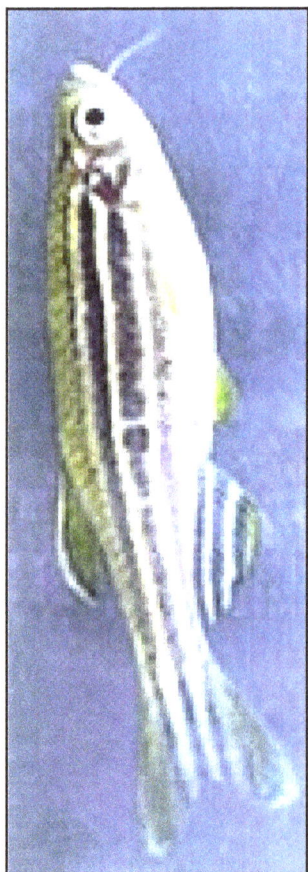

Figure 7.JD.1: *Danio malbaricus* (Page 153)

Figure 7.JD.2: *Danio albolineata* (Page 154)

Figure 7.JD.4a: *Zebra danio* (Page 156)

Figure 7.JD.3: *Spotted danio* (Page 155)

Figure 7.JD.4b: *Brachydanio rerio*
(Page 156)

**Figure 7.JG.1.1: Shubunkin Goldfish
(Other variety) (Page 159)**

Figure 7.JG.1: Gold Fish Shubunkin (Page 159)

Figure 7.JG.2: Comet Gold Fish (Page 160)

**Figure 7.JG.3: Fantail Goldfish
(Page 160)**

Figure 7.JG.4: Veil Tail Goldfish (Page 160)

Figure 7.JG.5: Pearl Scale Gold Fish (Page 161)

Figure 7.JG.6: Black Moor Telescope (Page 161)

Figure 7.JG.7: Celestial Bubble Eye Goldfish (Page 161)

Figure 7.JG.8: Black Oranada (Page 162)

Figure 7.JG.9: Lion Head Goldfish (Page 162)

Figure 7.JG.10: Teleoscopic Eyes (Page 163)

Figure 7.JG.11: Goldfish Jikin (Page 163)

Figure 7.JG.12: Tosakin Goldfish (Page 164)

Varieties of Gold Fish

Calico Gold (Page 165)

Common White Goldfish (Page 165)

Lion Head Gold Fish (Page 165)

Red Cap Goldfish (Page 166)

Red and White Goldfish (Page 166)

Red and White Butterfly Tailed Goldfish (Page 166)

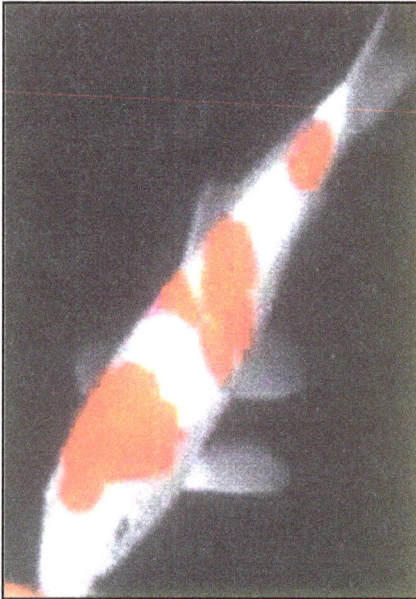

Figure 7.JG.13: Koi (Page 172)

Figure 7.JR.1: Bottom: Three Strip *M. Trifasciata*
Figure 7.JR.2: Top: *M. arfakensis* (Page 174)

Figure 7.JR.3: *Rasbora maculate* (Page 175)

Figure 7.JR.4: *Rasbora heteromorpha* (Page 175)

Figure 7.JR.5: *Rasbora pauiperforata* (Page 176)

Figure 7.JR.6: *Rasbora trilineata* (Page 176)

Figure 7.JR.7: *Rasbora parluciosoma libosa* (Page 176)

Figure 7.K.1: *Aplochelius Aplochelius lineatus* (Stripped Panchax) (Page 178)

19

Figure 7.K.2: *Aplochelius blockii*
(Raibow Panchax) (Page 179)

Figure 7.K.3: *Aplochelius panchax*
(Blue Panchax) (Page 179)

Figure 7.K.4: *A. Dayi*
(Green Panchax) (Page 179)

Figure 7.L.1: *Macrognathus
aculeatus* (Page 181)

Figure 7.L.2: *Mastacembalus armatus* (Page 182)

Figure 7.M.1: *Badis
badis badis* (Page 183)

Figure 7.M.2: *Badis badis salnensis* (Page 183)

Figure 7.M.3: *Badis badis burmanicus* (Page 184)

Figure 7.N.1: Black Molly (Page 185)

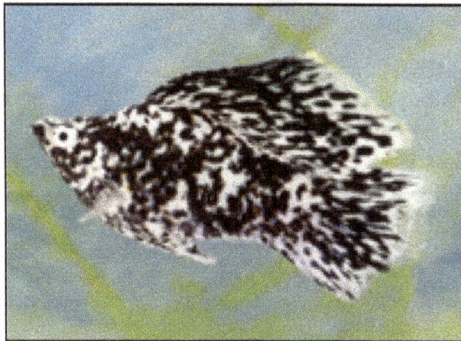

Figure 7.N.2: Marbel Molly (*Poicelia latipinna*) (Page 187)

Figure 7.N.3: Golden Ballon (*Xiphophotus variatus*) (Page 187)

Figure 7.N.4: Sward Tail (*Xiphophorus helleri*) (Page 188)

Figure 7.N.5: Platy (M and F) (Page 190)

Varieties of Guppy

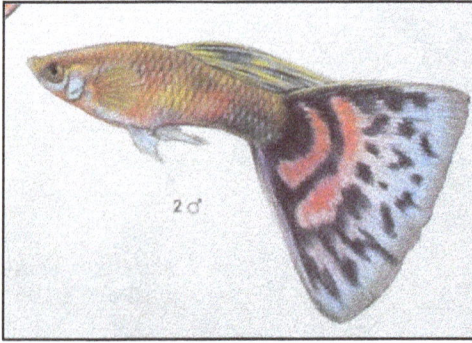

Guppy (Male) (Page 193)

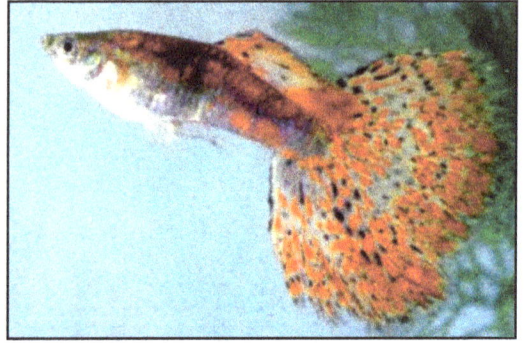

Delicate Variegated Guppy (Page 193)

Black Guppy (Page 193)

Green Variegated Guppy (Page 193)

Blue Variegated Guppy (Page 194)

Red Snakeskin Guppy (Page 194)

Green Snakeskin Guppy (Page 194)

Tuxedo Cobra (Dragon Head) Guppy (Page 195)

Tuxedo Guppy (Page 195)

Tuxedo Cobra (Dragon Head) Guppy (Page 195)

Red Tail Guppy (Page 196)

Blond Red Tail Guppy (Page 196)

Blue Tail Guppy (Page 196)

Pineapple Guppy (Page 197)

Electric Blue Guppy (Page 197)

23

Double Sword Guppy (Page 197)

Leopard Guppy (Page 198)

Lyretail Multicolor Guppy (Page 198)

Figure 7.P.1: Oscar (Page 201)

Figure 7.O.2: *Gambusia affinis affinis* (Page 199)

Figure 7.Q.1: *Monotetrus travancoricus* (Page 203)

Figure 7.Q.2: *Tetradon fluviatilus* (Page 203)

Figure 7.Q.4: *Tetradon mbu* (Page 204)

Figure 7.Q.3: *Tetradon nigroviridis* (Page 204)

Figure 7.R.1: *Tilapia mosambica* (Page 206)

Figure 7.S.1: *Labio bicolor* (Page 207)

Figure 8.1: *Acanthus leacosternon* (Page 215)

Figure 8.2: *Acanthurus glaucopairies* (Page 216)

Figure 8.3: *Acanthurus sohal* (Page 216)

Figure 8.4: Marine Clown Fish (Page 217)

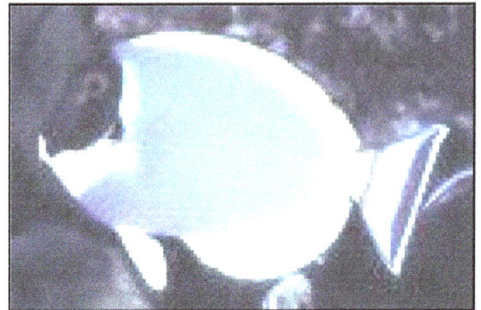

Figure 8.5: *Acanthurus lineatus* (Page 217)

Figure 8.6: Flying Fish (Page 218)

Figure 8.7: *Chaetodon frembii* (Page 219)

Figure 8.8: *Oxycirrhites types* (Page 220)

Figure 8.9: *Gobiosoma oceanops* (Page 221)

Figure 8.10: *Gobiodon okinawae* (Page 222)

Figure 8.13: *Opistognathus auriform* (Page 226)

Figure 8.11: *Goblodon citrinus* (Page 223)

Figure 8.14: *Amphirion clerkii* (Page 227)

Figure 8.12: *Pseudocheilinus hexataenia* (Page 225)

Figure 8.15: *Chromis caerulea* (Page 228)

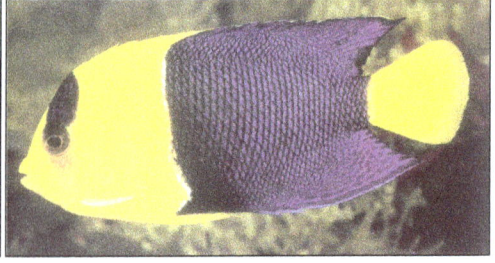
Figure 8.16: *Centropyge bicolor* (Page 229)

Figure 8.17: *Centropyge loriculus* (Page 230)

Figure 8.19: Marine Angel
(*P. semicirculatus*) (Page 231)

Figure 8.18: *Pomacanthus annularis* (Page 231)

Figure 8.20: *Hippocanpus erectus* (Page 232)

Figure 8.21: *Hippocanpus huda*
(Page 232)

Figure 8.22: Red Arowana (Page 233)

Figure 8.23: Golden Arowana (Page 234)

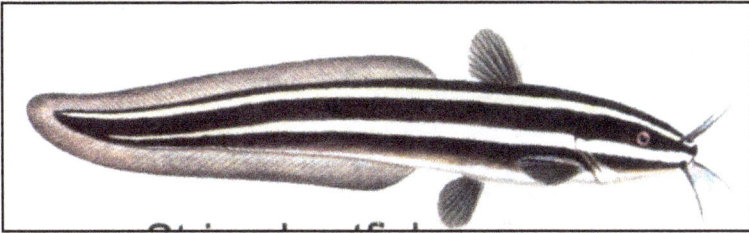

Figure 8.24: Striped Catfish (Page 235)

Figure 8.25: *Mystus vittatus* (Page 236)

Different Types of Fish Diseases

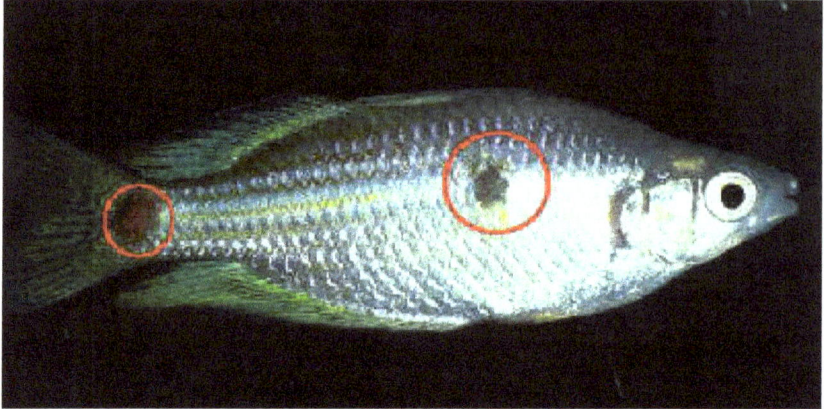

Figure 9.3: Tuberculosis with Raised Tumor Development (Page 244)

Figure 9.4: Black Widow Showing Typical "Ick" Sign, of a Folded Dorsal Fin, this Sign Often Appears in Early Stage (Page 245)

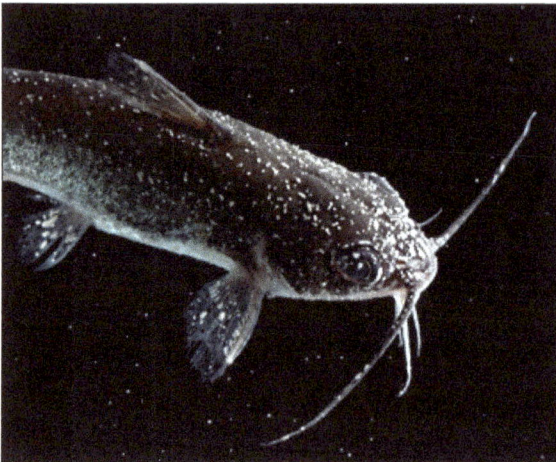

Figure 9.5: A Catfish Heavy Infested with White Spot (Page 245)

Figure 9.6: Discus Fish Infested by Costia (Page 246)

Figure 9.7: Fish Affected with *Lernaea*–Anchor Worm (Page 246)

Different Types of Parasites

Figure 9.8: Argulus (*Gyrodactylus*) (Page 250)

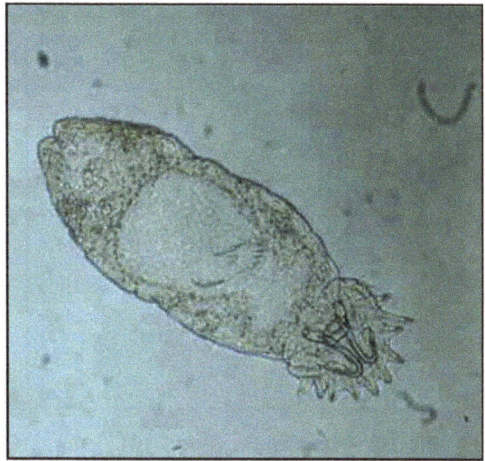

Figure 9.9: Skin Flukes (Page 250)

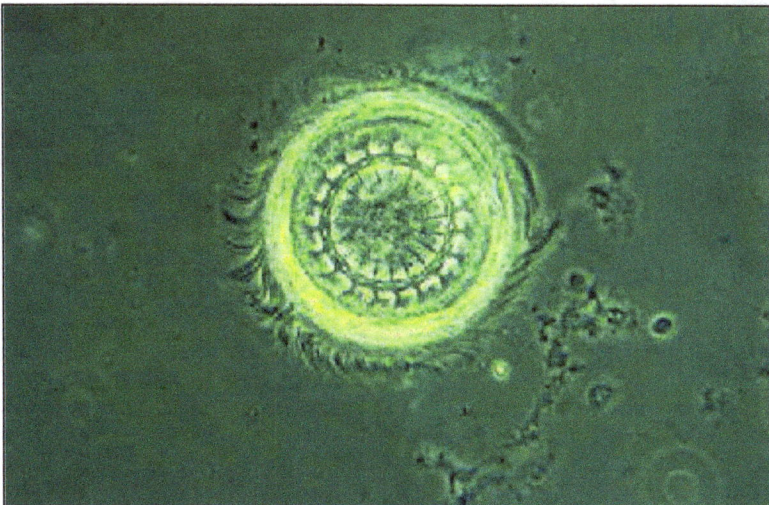

Figure 9.10: *Trichodina* (Slimy skin) (Page 250)

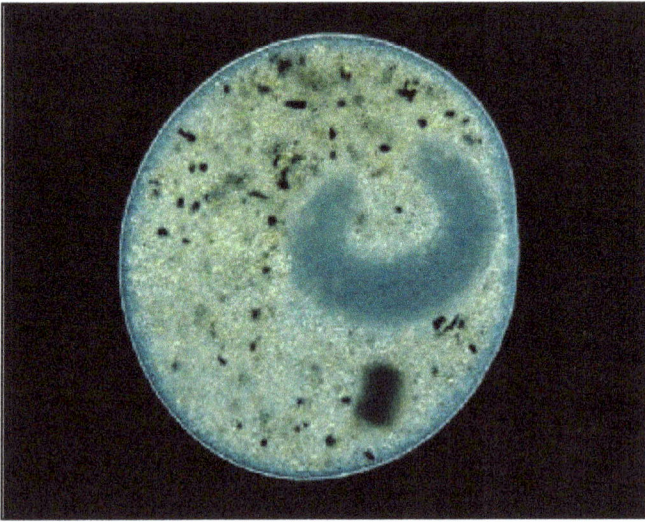

Figure 9.11: White Spot
(*Ichthyophthirius multifiliis*)
(Page 251)

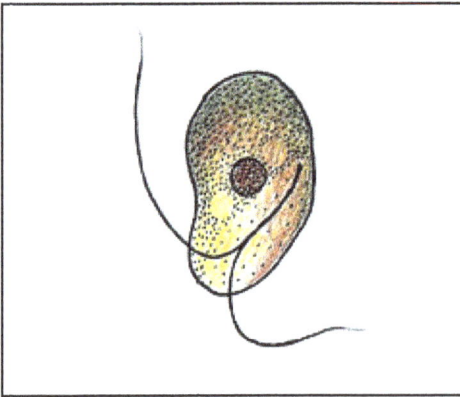

Figure 9.12: Costia (Page 251)

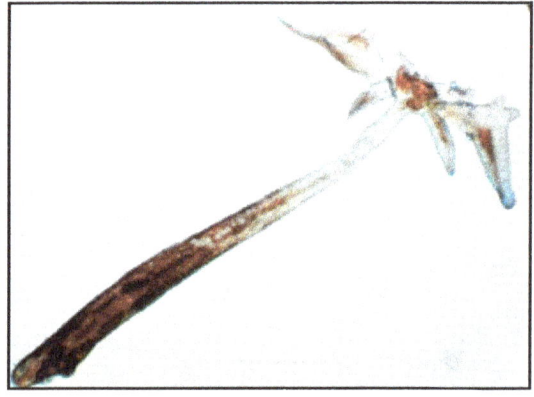

Figure 9.13: *Lernaea*–Anchor Worm (Page 251)

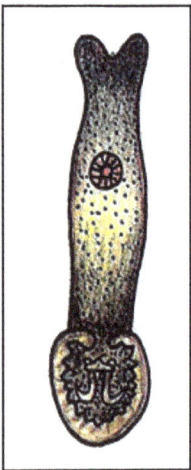

Figure 9.14:
Gyrodactylus–
The Gill Fluke
(Page 251)

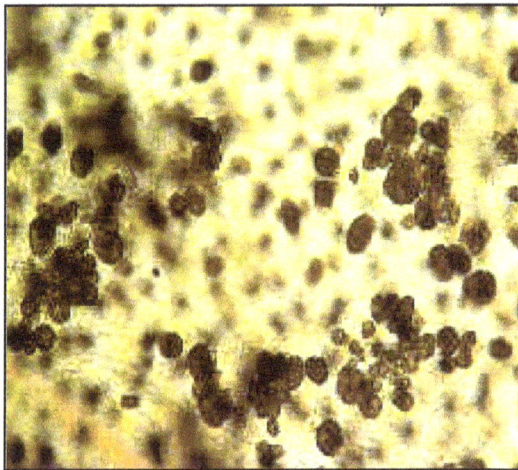

Figure 9.15: *Amyloodinium ocellatum*
(Page 251)

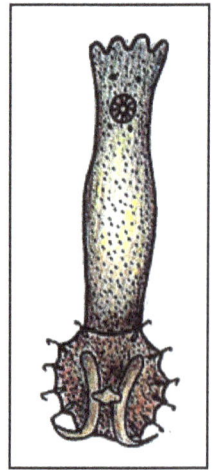

Figure 9.16:
Dactylogyrus–
The Gill Fluke
(Page 251)

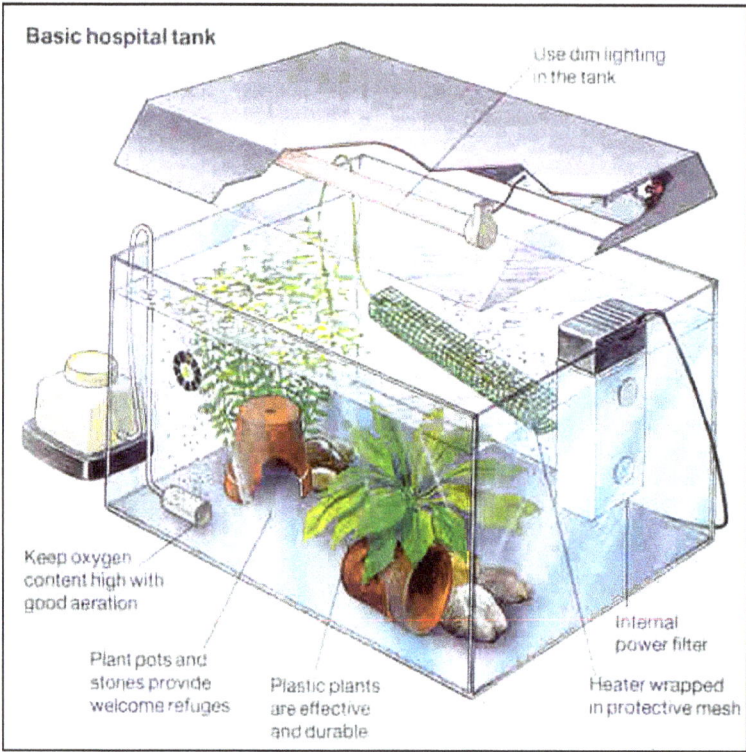

Figure 9.17: Basic Hospital Tank (Page 254)

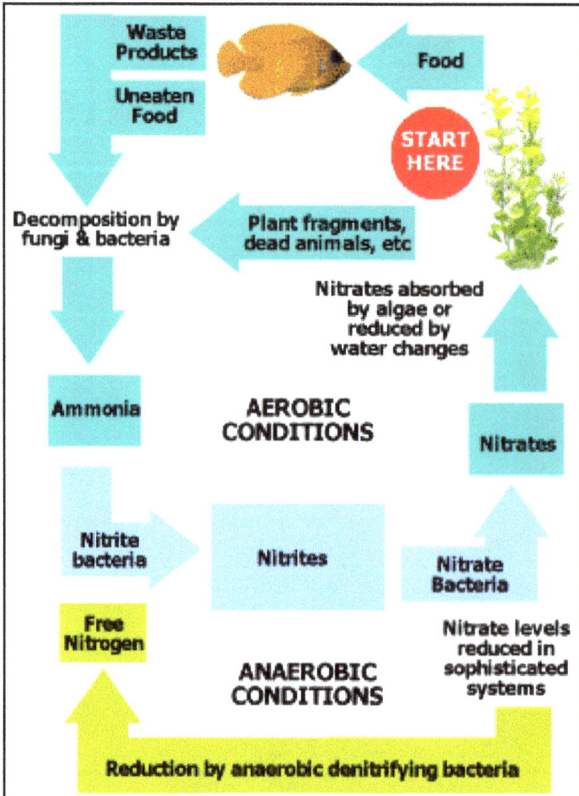

Utilization of Nitrates using Aerobic and Anaerobic Bacteria (Page 267)

www.ingramcontent.com/pod-product-compliance
Lightning Source LLC
Chambersburg PA
CBHW050508190326
41458CB00005B/1469